DOUBLEDAY
NEW YORK · LONDON · TORONTO · SYDNEY · AUCKLAND

NATURAL GRACE

Dialogues on

Creation, Darkness, and the

Soul in Spirituality and Science

MATTHEW FOX AND
RUPERT SHELDRAKE

PUBLISHED BY DOUBLEDAY

a division of Bantam Doubleday Dell Publishing Group, Inc.

1540 Broadway, New York, New York 10036

DOUBLEDAY and the portrayal of an anchor with a dolphin are
trademarks of Doubleday, a division of Bantam Doubleday Dell
Publishing Group, Inc.

Library of Congress Cataloging-in-Publication Data

Fox, Matthew.
Natural grace : dialogues on creation, darkness, and the soul in spirituality and
science / Matthew Fox and Rupert Sheldrake. — 1st ed.
p.   cm.
1. Religion and science.   2. Spirituality.   3. Creation.   4. Soul.   5. Good and evil.
6. Grace (Theology)   I. Title.
BL240.2.S55   1996
291.1'75—dc20   95-49785
CIP

ISBN 0-385-48356-2

June 1996

First Edition

1   3   5   7   9   10   8   6   4   2

In Memory of Bede Griffiths, O.S.B.

CONTENTS

PREFACE

We both share an interest in going beyond the current limitations of institutional science and mechanistic religion, and we both believe that as a new millennium dawns, a new vision is needed which brings together science, spirituality, and a sense of the sacred. Their separation underlies our present crises of ecological devastation, despair, and disempowerment. How else can hope in a new sense of meaning be awakened if not by the coming together of those two powerful traditions that were rent asunder in the seventeenth century? We need a new cosmology that speaks to our hearts as well as our minds.

The dialogues in this book are preliminary explorations into this new territory. Our hope is that our efforts will assist others to go further and enjoy themselves as much.

These dialogues emerged as we spoke together in public and private settings in London and Sheffield, England; in Berkeley, California; and on Cortes Island, British Columbia, Canada. We are grateful to the various audiences in these places for their encouragement and for questions and comments that helped us to clarify our ideas.

We are also grateful to our mutual friend Father Bede Griffiths, who died in 1993 and in whose memory we dedicate this book. He inspired us and many other people by his curiosity and breadth of vision regarding science, theology, and spiritual traditions of East and West.

We begin by introducing ourselves and saying something

about our backgrounds. In the first chapter we summarize the present situations in science and religion as we see them. In subsequent chapters we take up some of the implications that flow from a recognition of living Nature and the ways of grace.

INTRODUCTIONS

MATTHEW FOX

I was born and raised in Madison, Wisconsin, in the American Midwest. Wisconsin is a very beautiful and earth-centered place to grow up in. From the time I was very young I had Native American dreams. The spirits of the Native Americans are still very alive in the land of Wisconsin and they were a big influence on me. When I was twelve I had polio and lost the use of my legs; the doctors couldn't tell me if I would walk again. And when my legs came back in a year or two, I was overwhelmed with gratitude to the Universe for something I'd taken for granted for the twelve years when my legs had worked. In my ordeal I learned a lesson about not taking for granted, and this is a very mystical lesson.

The mystic is one who does not take for granted even a breath—especially a breath. And at some time in life we all come to that point. Unfortunately for some it's not until the end of life that we learn not to take breath, legs, or any of our gifts for granted.

As a teenager I found life very mystical. I remember a special moment during that time in my life occurring as I was walking into the living room, and someone in my large family (I was one of seven children) was playing Beethoven's Seventh Symphony on the stereo. The beauty of it powerfully hit me, and my soul just wanted to dance.

I also remember a similar experience reading Shakespeare in high school. But above all was Tolstoy's *War and Peace,* which I read when I was sixteen. I told a friend of mine it "blew my soul right open" and I wanted to explore this experience. I didn't really have a word for it then, but it was mysticism. That's why I went into the Dominican order after high school and two years at college, to explore spiritual experience.

That was about 1960 and John XXIII was pope—quite a different human being from the fellow who's running things these days. John XXIII moved me. He moved me as a human being; he moved everybody I think who knew him or knew of him. It was to his vision that I committed myself, and I took my first vows as a Dominican when he was pope. Then of course there was the Second Vatican Council that he launched, which was very exciting for me as a young theology student. All these new ideas from people such as Teilhard de Chardin and others who had previously been condemned by the church for many years were now roaring into center stage. The sixties, the decade of my coming of age, was also characterized in my country by the civil rights movement in the early part and by the anti–Vietnam War movement in the latter part.

I found my life in the Dominican training very mystical, and yet we weren't studying the mystics. For example, in twelve years of training as a Dominican I never heard the name Meister Eckhart mentioned once. The greatest mystic of the West—and he was a Dominican, no less. We didn't have a single course on the mystics or on spirituality as such. You were meant to get it by osmosis. We practiced such things as vegetarianism, fasting, celibacy, and chanting, but there was no reflection on why or what it meant.

I urged my Dominican superiors that somebody should go and study spirituality. I told them that my generation was not interested in religion, we were interested in spirituality. The year I was ordained, I was allowed to go to Europe and get a doctorate in spirituality. I asked them, "Where do I go?" They said, "Go to Spain." I said, "No way. We don't need more Spanish spirituality." "Go to Rome," they said. "Oh no! Rome's the problem," I said. So I proposed that I write to Thomas Merton and ask him where to go. They thought that was kind of goofy, but I wrote to Merton and he said, "Go to Paris." So thanks to Merton I went to Paris and that's where I met my mentor, Père Chenu, a great French Dominican who used to come regularly to Ottawa, Canada, to teach and who is recognized as a grandfather of liberation theology.

It was he who identified by name the creation-centered spiritual tradition when I was studying with him in Paris. The fact that I found a mentor in him was very important. He cleared the way for me by naming this spiritual tradition of the West that combined mysticism and prophecy, spirituality, art, and social justice. For this was the single most pressing question I brought with me to Paris: How, if at all, can we relate spirituality and social justice? Gustavo Gutiérrez, the Latin American theologian who did groundbreaking work on liberation theology, cites Chenu often in his work. Chenu was a beautiful human being and incarnated the quest for justice with art and spirituality and with a wonderful sense of humor and great affection. He died at ninety-five on the day Mandela was released from jail, which I think was a perfect day for him to die. Chenu had spent twelve years in his own kind of prison, having been forbidden to publish by Pope Pius XII because of his contribution to the worker priest movement in France.

Returning to the United States, I started to teach and write and I got involved early in the seventies with the feminist movement. I taught in a women's college for four years and my experience of being there and hearing women's stories made me a feminist. I was reading a lot of feminist theology at the time—Mary Daly and Rosemary Ruether—and I thought that these people were asking the interesting questions. I was really bored with much of patriarchal theology—which one had all memorized by then. The feminists' questions were much closer to my questions about spirituality, such as images for God, symbols, and metaphors. They were getting deep into the soul issues that I found most patriarchal theology ignored.

Then the gay and lesbian liberation movement and the ecological movement emerged in the seventies. The question I've had for twenty-five years, and it shows in every one of my fifteen books, is How do you relate social justice to mysticism? Do they connect? I didn't find anyone connecting them in the sixties, except the cultural movement. But the cultural movement collapsed in both areas. In terms of mysticism, a lot of the counterculture went into drugs and got lost there. And in terms of social justice, a lot of energy went into anger, moral outrage, and not into creative, thoughtful, and effective social change.

So I *expect* the nineties to be regarded in time as a mature sixties. I hope *it will* be more mature both about mysticism and about the struggle for justice. When you bring the two together you have compassion: "Compassion is where peace and justice kiss," says the psalmist. And it seems to me that that's what all the mystical traditions were already telling us. We're divine, start acting divine—that means start acting compassionately. It's not very complicated.

By 1973 I realized that you can't teach spirituality in the Western model of education. This is why you don't get it in seminaries or any place else as such—our Western models of education, being Cartesian or Enlightenment oriented, are only about the left brain. Our education doesn't get to the heart that is in the body. It doesn't touch the body and it doesn't deal with the right brain, which is about awe and wonder.

Having come to this conclusion, I started a school called ICCS (the Institute in Culture and Creation Spirituality) in Chicago, where we bring together right and left brain, body work, native spiritualities, the new science, and a lot of art and meditation along with intellectual work. It works as a model of education, and now we're in Oakland, California, where we moved in 1982.

I wrote books over the years, and in the course of my work the Vatican objected to my teaching; in March 1993 I was expelled from the Dominican order of which I'd been a member for thirty-four years. But this is a policy in the present Vatican. It's getting to be less and less extraordinary to be attacked by this regime. Leonardo Boff, a friend of mine in Brazil who is the most-read liberation theologian in Latin America, had to leave his Franciscan order, and Eugene Dreuermann, the most-read Catholic theologian in Germany, was kicked out of the priesthood by his bishop. Six Spanish theologians were expelled from their diocese in Madrid recently. But they were taken in by a wonderful bishop, Bishop Casigalida, who works with the Amazon Indians in Brazil and whom I visited for a week a few years ago. He's now their bishop, and so they're still in Madrid doing their work.

I myself feel that the Vatican has made me a postdenominational priest in a postdenominational era. I think denominations are passé. People in their twenties today, the postmodern generation, don't even know the difference between a Lutheran and an Anglican and a Roman Catholic, much less care.

Now more than ever we have to strip down religions to their essence, which is not religion but spirituality. Spiritual experience must include worship that awakens people instead of bores them, that empowers them, that brings out the gifts of the community, that heals and brings the healers back to the center of the community, healers such as artists, justice-makers, and others. This is the agenda for the third millennium. Of course, it is creation's travail and pain, Gaia being crucified, that is the apocalyptic moment which calls us into new forms of spiritual expression, some of which are going to be ancient and some of which are going to be created by our generation out of necessity.

I've had the gift of working closely with scientists for the last fourteen years, especially Brian Swimme, a physicist. And Rupert and I have had dialogues that have been rich. Several things I really appreciate about Rupert are his simplicity, his poetry, and his knowledge of philosophy and theology. But also his courage. He's at least as crazy in science as I have been in theology. I think we share analogous wounds. I love the fact that he wrote his first book in Bede Griffiths's ashram in southern India and not at Cambridge whence he comes. I think it's part of his accomplishment as a human being and a scientist that he sticks his neck and soul out to admit, first, that he has a soul and, second, that inner work and professional work can actually go together and that it makes for a much more interesting way of doing science.

RUPERT SHELDRAKE

I was born and brought up in Newark-on-Trent, Notting-hamshire, in the English Midlands. My family were devout Methodists. I went to an Anglican boarding school. I was for a while torn between these two very different tradi-tions—one Protestant and the other Anglo-Catholic with in-cense and all the trappings of Catholicism.

But the thing that really preoccupied me was my interest in living things. From a very early age I was interested in plants and animals. My father was an amateur naturalist and microscopist and he encouraged this interest. My mother put up with it. I kept lots of animals at home and she said, as mothers always say, "It's all very well, but who's going to feed them?" And of course, in the end, she usually did.

I knew from quite an early age that I wanted to do biol-ogy, and I specialized in science at school. And then I went to Cambridge where I studied biology and biochemistry. However, as I proceeded in my studies, a great gulf opened between my original inspiration, namely an interest in life, actual living organisms, and the kind of biology I was taught: orthodox, mechanistic biology which essentially de-nies the life of organisms but instead treats them as ma-chines. And I had to learn that you can't respond emotion-ally to animals and plants. You can't connect with them in any way except by detached objective reason. There seemed to be very little connection between the direct experience of animals and plants and the way I was learning about them, manipulating them, dissecting them into smaller and smaller bits, getting down to the molecular level and seeing them as evolving by blind chance and blind forces of natural selection.

Well, I could learn this stuff; in fact, I was quite good at it. But the gulf grew bigger and bigger. When I was at Cambridge in the Biochemistry Department, I saw a wall chart showing the different chemical reactions in the body. And someone had written in big letters across the top of it *KNOW THYSELF*. This brought home to me a huge chasm between these enzymatic reactions and my own experience. The first thing we did in the Biochemistry Department was to kill the organisms we were studying and then grind them up to extract the DNA, the enzymes, and so on.

I felt more and more that there was something wrong, but I couldn't put my finger on it. No one else seemed to think there was anything wrong. Then a friend who was studying literature lent me a book on German philosophy containing an essay on the writings of Goethe, the poet and botanist. I discovered that Goethe at the end of the eighteenth century and beginning of the nineteenth had had a vision of a different kind of science, a holistic science that integrated direct experience and understanding. It didn't involve breaking everything down into pieces and denying the evidence of one's senses.

This filled me with great excitement, the idea that there could be a different kind of natural science. So invigorated was I by this prospect that I decided I wanted to study the history of science and philosophy to see why science had got to where it was. I was fortunate to get a fellowship at Harvard where I spent a year studying philosophy and history. Thomas Kuhn's book The Structure of Scientific Revolutions had recently come out and it had a big influence on me, gave me a new perspective. It made me realize that the mechanistic theory of life was what Kuhn called a paradigm, a collectively held model of reality, a belief system. He showed that periods of revolutionary change involved the replacement of old scientific paradigms by new ones. And if science had changed radically in the past, then perhaps it could change again in the future. I was very excited by that.

I had a wonderful time at Harvard. I soon discovered that in the United States, university students are treated like children—told exactly what to read and then tested to make sure they have read it. In England I hadn't been treated like that since I was fifteen. I didn't like that system at all. So I decided I didn't need the master's degree I was supposed to be getting. Anyway, I could simply buy one from Cambridge University. If you have a Cambridge B.A. you have to do only two things to get an M.A.: stay alive for three and a third years and save up five pounds, the price of the degree. The result was that I spent a wonderful year at Harvard freed from the tyranny of exams, tests, and so on. I could do exactly as I liked, go to any lectures in any subject, read anything. It was wonderful. Unfortunately, very few people have this experience at universities because they're nearly always on treadmills.

When I went back to Cambridge, England, I did a Ph.D. on how plants develop, particularly working on the hormones within plants. While I was a graduate student there I came across a group called the Epiphany Philosophers, who were connected with an Anglican monastery called the Community of the Epiphany. This group of philosophers, physicists, and mystics explored the connections among mystical experience, philosophy, and science—and still do. This was exactly what I was interested in, and I found this a very helpful and inspiring group.

However, this was a predominantly Christian group, and I wasn't a Christian. I was an atheist. When I was about fourteen, my biology master at school had convinced me that religion was a thing of the past, and science was the thing of the future. Religion shackled humans to superstition, priests, and dogma; but science liberated humans and enabled them to march forward to a new era of prosperity and brotherhood. Technological progress would bring about this kind of heaven on earth, through human reason, not through blind faith and mumbo jumbo.

Well, it was nice to think that I was in the vanguard of a heroic, liberating movement. I took an optimistic atheistic and humanistic attitude, which lasted a long time. It's a very firmly embedded mind-set once you get into it.

So when I joined the Epiphany Philosophers, the Christian aspect didn't interest me much. But I was interested in the exploration of new ideas in quantum theory, philosophy of science, parapsychology, alternative medicine, and the holistic philosophy of nature. These were some of the themes we were discussing in the sixties. We lived together as a community in an old windmill on the Norfolk coast for a week at a time, four times a year. It was an interesting mixture of people, ranging from undergraduates, hippies, and healers to eccentric professors, physicists, and monks — and we were all able to talk together and engage in common explorations.

I went on with my research on plant development and became a Research Fellow of Clare College in Cambridge and also a Research Fellow of the Royal Society, which gave me tremendous freedom, for which I'm very grateful. For seven years I lived in seventeenth-century rooms in a beautiful courtyard. I had all my meals provided. All I had to do was wait for a bell to ring and I just walked across the courtyard, put on my academic gown, and on high table was served delicious meals, with vintage wine from the well-stocked college cellars. After dinner we drank port in a paneled common room, called a combination room, and talked for hours. Since the fellows of colleges are from all different subjects. I had many valuable opportunities for interdisciplinary discussions.

I was free to do whatever research I liked. The first year, I went off to Malaya because I wanted to study plants in rainforests. I traveled through India and Sri Lanka on the way, and that was a real eye-opener. Being in Asia showed me totally different ways of looking at the world. When I got back to Cambridge, I went on with my work on plant development. As I did so I became more and more convinced that the mechanistic approach simply could not work in understanding the development of living organisms.

I was beginning to explore the holistic tradition in biology, which is a minority tradition, but it's always been there. I began to formulate the idea of morphic resonance, the basis of memory in nature, the main thing I've been working on since. The idea came to me in a moment of insight and was extremely exciting. It interested some of my colleagues at Clare College—philosophers, linguists, and classicists were quite open-minded. But the idea of mysterious telepathy-type interconnections between organisms and of collective memories within species didn't go down too well with my colleagues in the science labs. Not that they were aggressively hostile; they just made fun of it. Whenever I said something like "I've just got to go and make a telephone call," they said, "Ha ha, why bother? Do it by morphic resonance!"

I saw that a new kind of science was necessary, and I was encouraged as I began to see what it could look like. It became clear that the future of my interest didn't lie in biochemistry. I wanted to do something quite different, where I could work with whole organisms, and preferably do something that was useful as well. I resigned my fellowship at Cambridge and got a job in an international agricultural

*institute in southern India, at Hyderabad, where I worked
for about six years on the physiology of tropical legume
crops, improving crops for subsistence farmers in India.
This was a great opportunity to work in the fields, to get to
know plants year-round, growing outdoors, a completely
different experience from working with little bits of them in
the laboratory, where they are isolated from all the real-life
factors which are only too apparent in the world of agriculture.*

*The main reason for taking the job in Hyderabad was
that I wanted to be in India. I had already become interested in Indian philosophy and had started doing transcendental meditation. I was drawn to the Hindu traditions. So I
went to India, lived there, and worked in agricultural research. I loved being in India.*

*I went on thinking about my heretical ideas in biology
until I felt ready to write a book on the subject. I didn't
want to leave India but I had to leave my job because I was
working very long hours and didn't have time to write the
book. By that stage I'd met somebody who was to play a
great role in my life, Bede Griffiths, an English Benedictine
monk who lived in a small ashram right in very south of
India. It was a Christian ashram, bridging the Christian and
the Eastern traditions.*

*During my time in India I'd been involved with various
Hindu gurus and ashrams, and also with Sufis in Hyderabad, which has for centuries been a stronghold of Sufism. I
had a great friend who was a Sufi, an old and very charming
man, who was my teacher. But oddly enough, in spite of all
this, I found myself being drawn back to the Christian tradition, which I felt was my own tradition. I realized I could
never really become a Sufi because you have to become a
Muslim to become a real Sufi, and I couldn't see myself
getting into all that. I couldn't become a Hindu because I*

couldn't be an Indian. Yet at the same time I began to find a new meaning in the Christian tradition that I'd rejected for so long.

When I met Bede Griffiths, he made the bridge between the Christian and the Eastern traditions much easier for me to cross. I went and lived in his ashram for a year and a half, and then I wrote my first book, A New Science of Life, which I dedicated to him.

Then I went on working part-time in my old job in India and the rest of the time pursued my interest in morphic resonance and holistic ideas in biology. And that's what I've been doing since. I met my wife, Jill, in India and we share many interests. We've explored some of my interests together. One of the things we do is come with our two sons to North America every year, an annual family migration.

I've written three more books: The Presence of the Past, which expands the idea of memory in nature; The Rebirth of Nature, which shows how we can once again think of nature as alive instead of inanimate and mechanical; and Seven Experiments That Could Change the World, which is subtitled A Do-It-Yourself Guide to Revolutionary Science. In the latter I focus on areas where simple, inexpensive research can make a big difference. I think that only by reempowering independent investigation in science can the spirit of inquiry be revitalized. Science in the past was done mainly by amateurs. It has now become exclusively professionalized, but it doesn't have to be that way.

One of my main concerns is the opening up of science. Another is exploring the connections between science and spirituality, and I am particularly thankful for the opportunity to do this with Matthew Fox.

CHAPTER ONE

LIVING NATURE AND CREATION SPIRITUALITY

RUPERT

In thinking about the relation of God and nature, much depends on how we conceive of nature. A change in our view of nature is currently coming about through science itself. We are living through a major period of change in science, a paradigm shift, from the idea of nature as inanimate and mechanical to a new understanding of nature as organic and alive. The God of a living world is very different from the God of a world machine.

For most of human history, the vast majority of human beings have taken it for granted that Nature is alive. In our own culture people habitually spoke of Mother Nature and Mother Earth. It was taken for granted that we live in a living world. This view was held by the Jews, among other ancient peoples, and also by the Greeks. The Greeks thought of the entire cosmos as a vast living organism, like a giant animal with a body, soul, and spirit. And everything within the world participated in this life of the cosmos. Theirs was what we'd now call a holistic vision. The Greeks made it more explicit than most, and it was inherited from the Greek thinkers of the Middle Ages, particularly through the philosophy of Aristotle.

The official doctrine in the Europe of the Middle Ages was animism—the belief that nature is alive, that the world is a living world. Animals and plants had souls—they were truly animate. The world was pervaded by all kinds of spiritual and psychic entities. It was a very different worldview from the one predominant today, and it had much in common with the animistic philosophies of all traditional peoples. The religion of the Middle Ages was Christian animism.

This animism not only was a sophisticated philosophy taught in the medieval universities but also was reinforced by popular practices, many of which were pre-Christian traditions that continued to exist in Christian form, such as a recognition of the Holy Mother, the honoring of holy wells and springs, pilgrimages to places of power, and the celebration of seasonal festivals linking the human community to the cycles of nature.

In northern Europe, much changed at the Protestant Reformation when shrines, pilgrimages, the cult of the Holy Mother, and many of the great festivals were suppressed as relics of paganism, which indeed they were. In the Protestant countries these changes effectively desacralized the natural world. Religion then became almost exclusively preoccupied with the interplay of humans and God, the drama of fall and redemption. The natural world was devoid of spiritual power, at best a neutral backdrop.

Desacralized nature could then be seen in a new way. There were no longer any religious restraints to the con-

quest and exploitation of Nature. Everything was up for grabs. It was Sir Francis Bacon who in the early seventeenth century most clearly put forward a new agenda for human domination through science and technology, emphasizing the need for organized empirical research. Through probing Nature's secret places, as he put it, man could find out her secrets so that he could more effectively bind her into servitude and have dominion and power over her.

The feminists have had a good time with this kind of language, as you can imagine. Mother Nature no longer had any intrinsic value of her own, but was simply there for man to use as he saw fit. Bacon helped prepare the way for the mechanistic revolution in science. This revolution first came to consciousness in a vision of René Descartes on November 10, 1619. Descartes claimed that this vision was given to him by the Angel of Truth; in modern parlance we would say that it was channeled. It was a vision of a machinelike world, governed entirely by universal mathematical laws, with no inherent spontaneity or freedom. This was the essence of the mechanistic theory of Nature.

The soul, the animating principle, was withdrawn from the whole of Nature and from the human body too. The world was deanimated and effectively became regarded as an automatic machine with no soul, no spontaneous life, and no purposes of its own. Animals and plants were thought of as inanimate machines, and so was the human body. The only part of the material world that was not entirely mechanical was a small region of the human brain, the pineal gland, where the rational conscious mind somehow interacted with the machinery of the nerves. The old view was not that the soul is in the body, but that the body is in the soul. Now the soul survived only inside human heads.

This desacralized, deanimated, soulless vision of Nature became the foundation for modern science and was estab-

lished as its reigning paradigm in the scientific revolution of the seventeenth century. The new science involved an explicit rejection of the traditional idea that Nature is animate and that all living beings have souls and purposes of their own. Mother Nature was now regarded as dead matter, subject only to mechanical forces and governed by mathematical laws.

I will now briefly go through the essential features of the mechanistic worldview, which is still the official philosophy that dominates science, medicine, and agriculture; is generally taken for granted in the media, in politics, and in education; and underlies the ideology of economic development and technological progress. I will first spell out the main features of the mechanistic view of Nature and then show how every single one of them has now been transcended or superseded by the advances of science itself, which is now leading us toward a postmechanistic worldview.

The philosopher of science Sir Karl Popper put it succinctly by saying that through modern physics materialism has transcended itself. This is a very important point, and one that is not widely realized. The image of science that most people have is at least fifty years out of date and often a hundred years out of date. There is no good reason, other than habit, why we should go on teaching an outmoded scientific ideology to children in schools. New scientific ideas take much longer to filter into general awareness than ideas in the arts or fashion or politics—decades rather than months. For example, the quantum revolution in physics occurred in 1927, but it wasn't until the late seventies that it became a topic that could be discussed in polite society in England, following the publication of Fritjof Capra's The Tao of Physics in 1974. There is now a torrent of popular books about quantum theory, but there has been an awfully long time lag. In the normal run of events, the scientific *changes I'm going to be talking about will become part of popular consciousness somewhere around 2030. That may be too late.*

The first feature of the mechanistic worldview is that it's based on the machine as its central image: the world as a machine; animals and plants as machines; human bodies as machines. The job of science is to find out their mechanisms. By contrast, all previous worldviews took organisms as their principal source of metaphors and myths. Mechanists dismiss such organic metaphors as subjective or anthropocentric. But ironically, the machine is one of the most anthropocentric of metaphors, for only people make machines, and only recent people at that. The mechanistic worldview involves projecting modern man's fascination with machines onto the whole of nature.

This mechanistic universe is inanimate and purposeless. Inanimate *literally means "soulless."* Purposeless *means "without any internal purposes."* The whole course of nature is supposed to be pushed by causes from behind rather than drawn by attractions or motivations from ahead, moving toward goals.

In the seventeenth century, matter became mere dead, unconscious stuff, made up of inert atoms. The Earth was thought of as a misty ball of rock hurtling around the sun in accordance with Newton's laws of motion; it had no life of its own.

The whole course of Nature was thought to be determined. Everything went on inexorably, mechanically, and was in principle completely predictable. The whole of nature was thought to be essentially knowable to the mathematical reason of scientists.

The kind of knowledge that scientists had of the world was essentially disembodied. It was as if the scientist was not involved in what he was doing, as if he was seeing the world from outside. This was an essential part of the Copernican revolution in the sixteenth century. The idea that the Earth was spinning around and was going around the sun, rather than the sun going around the Earth, involved seeing the Earth from outside, as it were. This is rather like the visionary journeying of shamans, traveling out of their bodies and looking at the Earth from outside. The globe in every school classroom is a familiar reminder of this vision of the Earth from outside. It wasn't until astronauts got up into space that this view of the world from outside was turned into an actual experience. This imaginative disembodying was essential to the scientific revolution.

The idea that scientists are somehow disembodied, not bodily or emotionally involved in what they're doing, is part of the style of science to this day. Scientific papers are full of language like "observations were made," and children learn to write in their school notebooks this so-called objective style: "a test tube was taken." No one actually does anything; things just happen in front of the observing, detached scientist. Of course, the reality of scientific research is very different, as anyone who has worked in a lab knows.

In the seventeenth century, scientists assimilated the idea of God into their mechanistic worldview. The whole of Nature was thought to be created by God, but to be in itself uncreative. It was governed by changeless laws that were thought of as eternal mathematical ideas in the mind of a mathematical God.

The God of the world machine was conceived of as a designing intelligence, engineer, and mathematician—a clear case of picturing God in the image of man. This shift in worldview to the mechanistic philosophy was adopted by many theologians. Newton and Descartes themselves were very interested in theology and laid the foundations of mechanistic theology. Protestant theologians in particular have tended to take it for granted that animals, plants, and the world as a whole are machines and have then tried to fit God into this mechanical picture as the great machine maker.

This view of God has committed a whole tradition of theology—natural theology—to a mechanistic worldview. Animals and plants were regarded as machines, and their beautiful adaptations were taken as evidence of intelligent design by an external, machine-making God.

This is essentially what the scientific revolution of the seventeenth century put in place, and it is still the basis of scientific ideology. The word mechanistic in most circles is used pejoratively, but in modern science, and especially in biology, it is a good word. The good guys are the mechanists; the bad guys are the vitalists or the animists. One has to remember that the sciences have developed unevenly and that the changes I'm going to talk about have happened mainly in physics. Academic biology and medicine are still under the sway of the mechanistic worldview, living fossils of an older mode of thought.

First of all, the idea of the cosmos as a machine has given way to the image of the cosmos as an organism. The Big Bang theory, which has been orthodox since around 1966, tells us that the Universe began small and has been growing ever since. As it grows, a succession of new structures and forms appears within it. This is nothing like any machine we know of. But it is like the way an embryo develops or a tree grows from seed. It implicitly means that cosmology has adopted an image of a developing organism as opposed to a machine (Figure 1.1).

Cosmologies tend to mirror social preoccupations, and in the 1960s it was assumed that the cosmos would grow indefinitely, in harmony with the idea that economies would grow indefinitely. We now have more doubts about unlim-

MECHANISTIC UNIVERSE	LIVING COSMOS
Machine	Developing organism
Inanimate	Fields
Purposeless	Attractors
Inert atoms	Structures of activity
Earth dead	Gaia
Determinate	Indeterminate, chaotic
Knowable	Dark matter
Disembodied knowledge	Participatory knowing
Uncreative	Creative evolution
Eternal laws	Habits

ited economic growth, and in cosmology, sure enough, there is now something called dark matter looming in the background. No one knows what it is, but if there is enough of it, it will cause the expansion to slow down, and eventually the universe will begin to contract until everything's annihilated in the opposite of the Big Bang—the Big Crunch.

The idea that Nature is inanimate has been replaced by the idea that Nature is organized by fields. Fields, like souls, are invisible organizing principles. Magnets in the ancient Greek period, and right up until the seventeenth century, were thought to have souls. The soul was the invisible thing in and around the magnet that was responsible for its attractive and repulsive powers. Now we think of magnets having fields.

In area after area of science, the old idea of the soul as an invisible organizing principle has been replaced by the concept of a field. So I would say that Nature is being reanimated through fields, which have taken on many of the traditional roles of souls in the premechanistic paradigm.

Souls motivated organisms by attraction. The soul of the oak tree, according to Aristotle and St. Thomas Aquinas, drew the oak tree toward its mature form. The growing seedling was attracted toward the mature form of the oak. This kind of motivation by attraction was dismissed from science in the seventeenth century. It has recently been smuggled back in through the concept of attractors. These are central to the modern science of dynamics and enable processes to be modeled in terms of where they're going to end up rather than in terms of how they're being pushed from behind. I can't go into the technical details of this now, but it is significant that this concept has become very influential within science.

The idea of inert atoms has given way to the idea of atoms as structures of activity. They do not consist of fixed, inert stuff, but rather of energy moving and oscillating within fields. So matter itself has turned out not to be fundamental; fields and energy are more fundamental.

The idea that the Earth is dead is giving way to the Gaia hypothesis. Gaia is the Greek name for Mother Earth. Modern science, following James Lovelock, is rediscovering the concept of the Earth as a living organism. This is news to some people in the West; but it's not news to most people throughout the world. I've tried talking to Indian villagers about this and they are profoundly unimpressed. When you tell them that modern science is now discovering that the Earth is a living organism called Mother Earth, you are telling them what they've known all along. And in a sense all of us have known it all along, but there's now a way in which this old idea can be formulated scientifically. The Gaia hypothesis is a major step toward a recovery of the sense of the living world.

The doctrine that everything is determinate, in principle totally predictable, suffered a blow with the development of quantum theory in the 1920s, when it was realized that there was indeterminism at the microscopic level. More recently the recognition of chaos and chaotic dynamics has made the old idea of determinism untenable not just in the quantum realm, but in the weather, in breaking waves, in the activity of the brain, and indeed in most natural systems. Even the dynamics of the solar system, the classic exemplar of Newtonian mechanics, turns out to be chaotic in the long term. So chaos and indeterminism have given us a greater sense of the freedom and spontaneity of Nature than anything that prevailed for more than three centuries, while science was under the spell of the fantasy of total predictability. The idea of the whole of Nature as totally knowable has also suffered a terrible blow with the discovery of dark matter. It now turns out that ninety to ninety-nine percent of matter in the Universe is utterly unknown to us. It's as if physics has discovered the cosmic unconscious. Dark matter determines the structure and fate of the Universe, and yet we haven't a clue what it is.

The idea of the disembodied knowledge of the scientist is giving way to a sense of science as participatory. The observer is involved in what he or she observes. What is being looked for, and the way it is looked for, affects what is found. Moreover, the expectations of the experimenter affect what is observed, as in the placebo effect in medicine or the experimenter effect in psychology. We are coming to a more participatory sense of our knowledge of Nature.

The idea of Nature as uncreative has been superseded by the idea of creative evolution. Darwin helped us to recognize that Nature herself is giving rise to new forms of life in the biological realm, and in the light of the Big Bang theory we now see the entire cosmos as a creative evolutionary system. This of course raises the question of the nature of creativity in a new way. In an evolutionary world, creativity is an ongoing feature of the developing cosmos.

And finally, eternal laws. These made sense in an eternal

world, but I don't think that eternal laws of Nature make much sense in an evolving world. Either one has to have an idea of evolving laws or, better, one has to get rid of this very anthropocentric metaphor. As C. S. Lewis once remarked, "To say that a stone falls to earth because it is obeying a law makes it a man and even a citizen."

I prefer the idea of Nature as governed by habits. This is the main thrust of my own scientific work. The habits are maintained by a process I call morphic resonance, the influence of like on like. For example, if rats learn a new trick in London, then rats everywhere should be able to learn the same thing quicker because the rats have learned it in London. The more that learn it, the easier should it become everywhere. Likewise, if a new chemical compound is crystallized in New York for the first time, the more this is done, the easier it should become for these crystals to form all around the world. If children learn to play a new video game in Japan, it should be easier for children to learn the same thing in other countries. These effects should happen even without any normal means of communication.

This hypothesis is of course controversial and is still being tested. Most of the results so far point toward these effects being real. Nature may well have an inherent memory rather than being governed by eternal laws.

Taken together, these changes lead to a tremendous shift in worldview. But we're not going back to the premechanistic kind of animism. We are now in a postmechanistic state, at a higher turn of the spiral, if you like. The new animism differs from the old animism in that living nature is now seen as developing and full of creativity. The premechanistic view was of the Universe as a mature organism, or even as a senescent and decadent organism. We now see it as an organism still growing and developing.

This new view of living Nature brings the need for a new

approach in theology. It makes a tremendous difference to the way that we think of the relations between God and Nature, because we've all grown up with an idea that nature is mechanistic, inanimate, and soulless. Most people have never been completely won over to the mechanistic philosophy but have tended to accept it from Monday to Friday. On the weekend, a different view of Nature takes over as millions try to get back to Nature in a car. On weekends, on vacation, and in their retirement fantasies, many people try to connect with Nature in a different way, forming an I/ Thou relationship rather than the I/It relationship that characterizes the mechanistic attitude. The idea of Nature as alive has been preserved by the romantic poets and is in accordance with many people's direct, intuitive experience of the natural world. It is taken for granted by children. So it is by no means unfamiliar; it has simply been relegated to the realm of subjective experience and private life, while mechanistic attitudes have been endowed with the authority of scientific objectivity. But now science itself is leading us toward a new understanding of the life of Nature.

MATTHEW

Just as Rupert has been working on changing the old view of science to something that works today, so I have found myself in the same work in our religious tradition in the West. I too suffered from our reductionist science when I was a first-year student in biology and I put the following question to my professor: "What is life?" And he said, "You can't ask that question in a biology class, only in philosophy courses." So that did it, and I got turned off by science.

I remember the year of spiritual training that began my admission as a Dominican. We had a wonderful place in the hills overlooking the Mississippi River, and I proposed to the community one beautiful spring day that we shouldn't just meditate in the chapel, we should go outdoors and meditate. They thought that was not appropriate and so I found time on my own to do that. Once again, I remember interacting with a bud coming out of a tree in springtime, and the same question kept coming up: "What is life?" And I realize when I look over my work in creation spirituality that this theme of the sacredness of life, the mystery of life, and the revelation of life keeps roaring back. For example, in my book *Musical, Mystical Bear*, which was my first real effort to probe the essence of spirituality, I asked the question "What is prayer?" And I came up finally with the proposal that prayer is a radical response to *life.* I remember when I first tried out those ideas on some fellow theologicans, one of them said, "Life? Life is so abstract," whereas more recently, in reading the mystics Hildegard of Bingen and Thomas Aquinas, I notice that they both say, "God is Life."

It takes the mystic inside every one of us, it takes the child, the vulnerable child wanting to play in the Universe, to respond playfully and pleasurably to life. That's what mysticism is. Essentially it's the pleasure principle, it's the "yes" that we make to life.

But the other radical response is that of the prophet in us, which is the "no" that we make in life. As Rupert is standing up and saying "no" to a mechanistic system that has overtaken our science and culture, so the prophet always stands up, as Rabbi Abraham Heschel says, to interfere with the forces of injustice, to take a stand. I played with that dialectic of mysticism and prophecy, yes and no, as being the core energy of the spiritual journey.

In what Rupert has just said, I am struck with many connections. One is his point that the last time we had an animistic Christianity in the West was the Middle Ages.

That is important. I have been drawn to the Middle Ages in my efforts to find a spirituality that speaks to the deep needs of our time. Partly of course it's my Dominican training because Dominic and Aquinas and Meister Eckhart were all part of this medieval renaissance that you can still see in Europe in cathedrals such as Lincoln and Chartres.

This is the sense of the cosmic mother and the goddess present in all Nature, which is also the sense of the cosmic Christ. Rupert reminded me of a line from Thomas Aquinas, who said, "A mistake about creation results in a mistake about God."

All these mistakes on the left-hand side of the chart that Rupert has made were also accepted by the theological, academic, religious, and seminarian establishments of the last three hundred years. If you line up those left-hand sides — machine, inanimate, purposeless, inert, Earth-dead, determined, uncreative, and eternal laws — that almost sounds to me like a description of ninety-five percent of the worship that is going on in the West today.

"A mistake about creation results in a mistake about God." The paradigm shift that Rupert and others are naming can ignite a spiritual breakthrough, a moment of spiritual renaissance, a rebirth of divinity. Back in the late sixties, we said that God is dead, meaning of course that we had killed God. Now we're saying that the planet is dead, meaning of course that we're busy killing it. And I propose that worship is dead.

But the issue of course isn't God being dead, Earth being dead, or worship being dead. It's us. It's our souls, as Rupert was saying. Can we come alive again? Is there resurrection? Thomas Aquinas said there are two resurrections, and the first is waking up in this lifetime. He applies all the resurrection texts, for example, Paul talking about the trumpet that will sound. Aquinas said this applies to our waking up now. If we would respond to this first resurrection, the second one after this life passes on will take care of itself. Aquinas also said that revelation comes in two volumes, that of the Bible and that of Nature.

It is that dimension of the revelation of nature that has been lost in the last four hundred years in the West. Rupert is quite hard on the Reformation and on Protestantism in his book *The Rebirth of Nature,* and I think a point underlying his criticism is this: Protestantism was indeed a prophetic movement about a corrupt church situation, but its strengths and its weaknesses are in the same place. It emerged at the time of the printing press, at the time of the rediscovery, the remaking of the written word. And in many ways religion and theology in the West went the way of the head, the left hemisphere of the brain: reading, mechanically dissecting the texts. It was thought that somehow there was salvation in getting the right texts, which is really quite fool-ish because the Bible was not first written at all. The people who proposed the Bible were illiterate and certainly didn't know seven biblical languages in order to create it.

Western religion has been stuck in verbal theology for four hundred years and it fits the basic academic model that we have taken uncritically from Descartes and Kant. This is very patriarchal, the idea that truth is found somehow from the neck up.

And so this challenge from Aquinas. Two volumes, Na-ture and the Bible. To listen to Rupert and other scientists of our time is to pay attention to the revelation of Nature. This leads to a breakthrough into other modes of doing our spiritual practice and our theology. Aquinas said, "God is per se life." God is the most living of all beings and shares this life participation with all beings. Aquinas said, "Every creature shares in the dignity of causality, that God has shared with all creatures." Every time I use that phrase, "the dignity of causality," my back gets straight. You feel proud to be here. Every blade of grass shares in a dignity of causality: it can cause other blades of grass; it can cause cows to make milk and be happy and content. Every galaxy and supernova shares in the dignity of causality.

Then we also have a powerful tradition carried on by Meister Eckhart, who said, "Every creature is a word of God and a book about God." We must have the heart to open up and the willingness to be silent, to take in the wisdom of creatures and the wisdom of this Universe. This balance between the revelation from Scripture and the revelation from Nature is something that our generation yearns for.

There is a shift we see in theology today from theism to panentheism. When you look at the left hand of Rupert's chart, we're talking about a God who is above things and beyond things. Newton more or less pictured God as having an oil can behind the machine of the Universe to come in and keep it going once in a while. That's theism: we're here and God's out there someplace. Panentheism is the mystical tradition. Panentheism teaches that all things are in God, God is in all things, and God works through all things. We need to make this shift from theism to panentheism today in all our spiritual practice, in all our theology. And it can't be done even by theologians who've been around for years in academia. They too have to undergo a conversion from a theistic ideology that parallels the machine universe but does not speak to the spirit.

The West has not developed a theology of the spirit. We talk about a trinitarian divinity, but in fact in an excessively redemption-oriented and Jesus-oriented theology we leave out God the Creator who is per se life, who is present in all living things and therefore is understood more deeply when we understand the mystery of the living things. God the

Creator! The first article of faith has been often left behind in our theological training, as has the third article of faith, which is about God the Spirit, sanctification, our divinization. How many of us, if we practice religion in the West at all, have ever heard sermons on our divinity, what it means to be divine. Yet Thomas Aquinas said, "God became human in order that human beings might become divine." Clement of Alexandria said the same thing: "The logos has become human so that humans may know what it means to be divine."

An understanding of our divinization, our sanctification, has not been developed and we haven't had a theology of the spirit. This is why there are fundamentalists saying that the issue is whether you believe in Jesus or not. I am a trinitarian Christian and that means experiencing the God of creation and the God of our divinization, God the Spirit. What is our Divinization? It is our creativity and our capacity for using this creativity for compassion. Being images of God means we are capable of a godlike kind of creativity and a godlike kind of compassion (which is healing by way of justice-making *and* celebration).

And so the issue of Jesus as Redeemer is not really enough. We must include the cosmic Christ. Because the cosmic Christ is that cosmic wisdom—Sophia—that permeates all creatures and the whole Universe. As the Book of Wisdom says, "It holds all things together in the Universe." The first hymns in the Christian liturgy in the first century were not hymns about our personal salvation. Not one of them talks about being saved and about what wretches they were until they met Jesus. They are all, as the Epistle to the Colossians says, about the Christ who binds all things together in heaven and on Earth and everything in between. It was a cosmic vision that the first response to the Jesus event provoked. And as Joseph Sittler, the Lutheran theologian, has said, this blaze of cosmic Christ theology has been turned down to a flickering flame in the West.

I'm not putting down the historical Jesus, but what I'm pointing out is that during the Enlightenment era our theological enterprise had put all of its eggs into the basket of the historical Jesus and deposed the cosmic Christ. Because the cosmic Christ represented the right column of Rupert's chart, it wasn't respectable to be mystical during a machine era. You don't want mystics in a machine; they'll foul it up for sure. You don't want children in a machine because a child wants to play in the pastures, in the fields of the Universe like a lamb; as the Book of Revelation says, they crucified the cosmic lamb. Why not the cosmic sheep? Christ was lamb because Christ represents the *puer* and the *puella,* the divine child wanting to play in every one of us.

We are crucifying the Christ all over again when we oppress God's creation, which is God's garden, which is God. It's an expression of the divine one, the cosmic Christ. God is not out there appointing us stewards of creation as if we're here to do the dirty work of an absentee landlord. Rather, divinity is in creation. When creation flourishes and is radiant, divinity flourishes. When creation is crucified, the Christ dies all over again.

I urge a return to all three articles of faith: creation, redemption (or liberation), and divinization or sanctification. And I call for an embrace of all three members of the Trinity: Creator, Son, and Spirit. And I call for our recovery of the cosmic Christ tradition to balance this excessive commitment to the historical Jesus because that era has ended. The quest for the historical Jesus of the last few hundred years found many things which are useful for us, such as how ancient the parables are straight from Jesus' mouth and how much trust the early Christian community had in its mystical experience of the Christ—so much trust that it did not hesitate to put many words into Jesus' mouth in the Gospel stories. But the era of the quest for the historical Jesus has ended. We need now to recover our cosmic Christ tradition. Thus we balance the prophetic Jesus with the mystical Christ.

A group of academic theological scholars in America called the Jesus Seminar made headlines around the country in 1992. They said that eighty percent of Jesus' words in the New Testament are not Jesus' words and that none of the words of John's Gospel belong to Jesus. A TV network came to California to interview theologians about this and while many agreed with the Jesus Seminar, they didn't want to be interviewed. They feared that they would get in trouble with the Vatican or something. I agreed to the interview. My first response to this great insight was to raise the question whether anyone ever said that every word in the Gospel is from Jesus' mouth. That's not the point! The *Christ* is just as important as Jesus is in the New Testament. John's Gospel is all about the Christ. And Paul never met Jesus. The first name given Jesus in the New Testament was given by Paul because his letters were written before the Gospels, and this name is Sophia—Lady Wisdom.

Lady Wisdom is walking around in the person of Jesus. You think that didn't shake up the Roman Empire and the synagogue? It's so radical an idea that our patriarchal theologians have not had it in their vocabulary for two hundred years. The first chapter of the Gospel according to John is not primarily about *logos;* that's the second generation. The matrix for John 1 comes straight from the Book of Sirach, which is about Sophia, Wisdom, searching to set up her tent in our midst.

Theology itself, religion itself, has to let go of religion. We have to let go of our theological modes of thinking of the last three or four hundred years; we have to get over our bias against mysticism if we are to be part of the solution and not the problem to the ecological crisis, the crisis of despair among the young and, as Rupert puts it so eloquently, the soulless, the deanimated soul of our own species.

In one of our conversations Bede Griffiths said this: "If Christianity cannot recover its mystical tradition and teach it, it should just fold up and go out of business; it has nothing to say." To that I say, Amen.

It is only out of the mystical tradition that our prophetic work to transform society will be authentically radical, will be rooted. Authentic action comes out of nonaction. It comes out of respect for mystery and wonder and the gift and the glorious surprise of our being here. And this is why I think the movement to a healthy, Earth-conscious, justice-oriented mysticism helps name in theological terms some of the themes that Rupert has named on the right side of his chart.

I often reflect on one of Rupert's contributions that has tremendous implications for theology and for our spiritual practice: his insight and insistence that the proper name for *soul* might be *field*. The question of what is soul has preoccupied me for a long time. Several years ago I read a book by the philosopher Charles Fair, who said that when a civilization loses its meaning of soul, it is coming to an end. I feel that our civilization lost its meaning of soul some time ago. We don't use the word with much assurance.

But of course that's the bad news. The good news is that when we can come to new images of soul, we're launching a new civilization. Now one thing I did in my pursuit was to go back to the mystics. I did a study of fifty-one images of soul in Meister Eckhart and fifty-one images of soul in Teresa of Ávila. The mystics are dealing with images, and that's how you redeem language—you go back to experience. So mysticism is about trusting your experience. And

the nearest thing to experience is not words, it is images. We
birth images out of our experience; the poets come along
and redeem the language for us, that's the process. But I feel
all this is happening in Rupert's work, especially in this
tantalizing image of our souls as fields.

I would like to offer a brief meditation, a response to his
proposal that soul might be a field. As I mentioned, I grew
up in Wisconsin, an agricultural state with a lot of rolling
fields and hills and lakes. Sometimes a field has snow on it
and it's very quiet. Sometimes it's very green and lush and
vibrant. Fields come plowed and they come unplowed.
Fields introduce us to horizons. Sometimes you can stand in
a field and feel there is no end or that there's an end that is
beckoning us and we call that a horizon. And then you get
there and there's still more; there's something unending
about a field, perhaps. Fields are earthy. Rupert says,
"Fields are more fundamental than matter."

Jesus talks about sowing seeds in a field. In fact, Meister
Eckhart at one point talks like Rupert when he says, "The
soul is a field." Jesus said, Sometimes the seed falls on rocks
or in brambles and does not grow; at other times it falls on
some good soil and flourishes. How do we respond, how do
we interact with the fields we find in one another? For example, when people are in love, are they establishing a new
field of relationship? Meister Eckhart says, "When I'm sitting here in Germany"—where he was sitting—"I can be
closer to someone I love who is in Jerusalem than I am to
someone right next to me here."

Field introduces a notion of space and time that is very
different from our literal, more Newtonian sense. Fields can
come flat, they can come curved—full of potential. And
fields are so deep that we don't even know where the field
ends and the Earth begins again in new form. Field is long
in developing. My understanding is that it takes ten thou-

sand years for God and Nature to grow one inch of topsoil. There's a patience to a field. Fields are also invisible, and these are the fields I think Rupert most likes to talk about: these invisible fields that we enter, the sacred spaces. What about the power at Chartres that keeps drawing people? The attraction of the field that's there and other sacred spaces in our lives. Essentially, a field, it seems to me, is a place in which we are called to be rooted. Back to the sense of roots and radical again, being capable of rootedness.

So I think that the term *field* connects psyche and cosmos together again. Paul Ricoeur, the French philosopher, says that psyche and cosmos are the same thing. In other words, the world we live in, the world we choose to live in, is our soul. So the fields we choose to play in, to struggle in, to redeem, these are our soul. As Rupert said earlier, and he's quoting a very ancient tradition, our souls are not in our bodies, our bodies are in our souls. Aquinas said that, Hildegard of Bingen said that, and Meister Eckhart said that. They may have been the last in the West to say that our bodies are in the soul. Maybe we should start saying our bodies are in the fields. What other images are awakened in us in hearing this metaphor of soul as field? What about the notion that divinity permeates the field? The pasture of the divine is, as Jesus said, among us: "The kingdom of God is among us." The fields are among us. People like Rupert perhaps are helping us become more aware of the implications of this.

The modern field concept has replaced the old soul concept as a matter of historical fact. The anima mundi, *the soul of the Universe, which according to the premechanistic cosmology held the stars in place and was responsible for the movements of the planets, has now been replaced by the universal gravitational field as the all-containing, invisible* organizing principle. Electromagnetic fields have replaced the old electrical and magnetic souls. In biology, morphogenetic fields, the organizing formative fields of the embryo and of the developing organism, have replaced the vegetative soul of Aristotle and the medieval philosophers.

I think the concept of fields helps to illuminate or even demystify some aspects of the older use of the word soul. It's very difficult for us nowadays to recover a sense of what people used to mean by soul. On the other hand, fields as usually conceived of in a mechanistic spirit are regarded as entirely unconscious. The challenge is to see how fields can be related to consciousness. In the old view of souls, most souls were unconscious. The animal and vegetative aspects of the human soul were not conscious. The old view of the psyche didn't require it to be conscious. We're trapped today because we have such limited words: if we say mind, we imply consciousness.

Yet many people since Descartes, notably Freud and Jung, have pointed out that much of our mind is unconscious. We've had to rediscover the unconscious aspects of the psyche. Because they are unconscious we don't know where they begin and end. Is my unconscious psyche confined to my own mind, or does it also connect me with all other people in the collective unconscious, as Jung would say? How is the collective unconscious related to the souls or fields of other animals, plants, Gaia, and the cosmos? We don't know.

One point that strikes me on the theological side concerns the nature of energy and fields. Nature as it is currently portrayed in science consists of energy, which can take any form, and fields, which are the formative principles in all realms of Nature. If we wanted to map them onto theological concepts, it is relatively easy in Eastern traditions. For example, in Hindu Tantrism, Shakti is the energetic princi-

ple, the moving, dynamic principle, and Shiva is the forma-
tive principle. In this case, Shakti is feminine and Shiva
masculine. It seems to me that in the Christian tradition the
energetic principle would be the spirit, which is always por-
trayed in terms of movement, change, the blowing wind, the
breath, the fire, the flying bird—these are all moving im-
ages. And the field would be the formative principle, the
logos, the formative aspect of creation. Have you ever
thought of fields in these terms?

Well, I have in fact, just before you suggested that spirit
would be the right metaphor. Theologically that's exactly
what I was thinking. Again, as I said earlier, we have a very
underdeveloped theology of spirit in the West. And part of
our theology of spirit comes from John, who says, "The
spirit blows where it will" (John 3:8)—the chance, chaotic
dimension of surprise that you alluded to earlier. But for me
the contemporary word for transcendence is not *up*—there
is no up in a curved universe, Buckminster Fuller points
out. There's only in and out—and for me the word *transcen-*
dence means *surprise.* And that's what the spirit brings.
And the spirit flows through all religions, all cultures, and
all of us. It's not easily controlled. An awakening of the
spirit is something our ecclesiastical bureaucracies are a little afraid of. But I think it's what a renaissance is all about.

The spirit, then, is a source of creativity. By contrast, the
morphic fields that I'm talking about are essentially habit-
ual. They are formative principles based on what's hap-
pened before. They are the carriers of habit, the carriers of
inheritance, good and bad. The way I see the Universe in
terms of modern science is of an interplay between the habit
principle, represented by fields, and the creative principle,
represented through the ongoing flux of things that's always

bringing about surprises. Indeed, our own lives seem to be an interplay of habit and creativity. Perhaps this fits in with what you've just said.

Very much so. Another word that we use for *habit* is *tradition*. But there's a danger of spelling it with a capital *T* instead of a small *t*. I was in Cardiff speaking on Celtic spirituality in a large Protestant church, and after the event a young man in his twenties came up and said, "You know, I go to church here and it's the first time I've ever seen the place filled. About thirty people come here on Sunday and it's so boring. Ritual's something they do to you. How can we reinvent ritual here?" I said, "You have to begin by removing all the benches and reimaging the field as curved because that's the kind of Universe we say we believe in. You need to get the breath moving because you can't pray without the body and the breath. The body and the heart go together." "Well," he said, "this church is 150 years old, it has a tradition and they wouldn't tolerate moving the benches." But I told him that the Universe is 15 billion years old and soon the human community has to make a decision: 150 years of human tradition or 15 billion years of Creation's tradition. "You may lose half your congregation, but you may gain a whole generation of worshipers who are eager to be invited into more curved spaces." This names the field and brings forth that morphic memory, because when we pray in circles, the circles of our ancestors and the circles of our curved Universe, we are arousing and awakening that memory and the resurrection that comes with that: a new life.

That reminds me of a point I want to ask you, arising from a question put to me last week by my three-year-old son. We were singing a song together about leaves in the spring and Mother Nature. He was fascinated by this and kept asking, "Who's Mother Nature?" I tried to explain. He went silent for a while and then asked, "Is Mother Nature Mary?" He knows about Mary because he has a picture of her in his room, and we quite often say the Hail Mary when he is going to bed. I paused for a moment and then said "yes," anticipating a theological revolution that may not have happened yet.

It seems to me that in many ways Mary has taken on many of the qualities of Mother Earth, especially when her shrines are in grottoes and caves and when she is associated with holy wells and springs. The actual practice of Marian devotion in many parts of the world suggests that she is a Christian form of the old mother goddess.

I know that the official theological line gives a much more guarded and limited conception of Mary. We need a fresh image of living Nature through the cosmic Christ, as you have shown so inspiringly in your books, but we also need a new sense of Mother Nature. In what form do you think this mother image or archetype is best conceived? Mary is not the only possibility. In some Greek icons, there are wonderful pictures of Mary with Jesus on her knee, and with Mary herself on the lap of her mother, St. Anne. Mary herself is seen against the background of the great mother, the grandmother. But St. Anne may not be the best candidate either. What do you think?

You're opening up a whole box of wonderful metaphors, and, as you say, like any series of metaphors no single one satisfies and that's what's wonderful—we keep going. There's a Black Madonna, well known in Europe, that is by all evidence a borrowing of Isis, the mother goddess of North Africa, and perhaps it also traces back to India and the mother goddess in India.

The fact that at the same moment of cosmological awakening in the twelvth century the West brought the Black
Madonna right into the heart of the cathedrals is very important. That word *cathedral* was invented in the twelfth
century. We know it's a Latin word for *throne* and we've
been taught that it's the place where the bishop puts his
bottom. But that is clericalism, it's patriarchy, and it's not
true. What *cathedral* meant in the twelfth century was the
throne where the mother goddess sits ruling the Universe
with wisdom and compassion and justice for the poor and
oppressed.

Whenever you rediscover the goddess, what you're really
rediscovering is the creative power of the community and of
the individual. The goddess tradition in Christianity is the
same as the cosmic Christ tradition; it's a synonym. Because
it's Sophia and it's cosmic wisdom: Lady Wisdom. The
depths of the Mary archetype are very powerful. They're
just beginning to emerge, and the wonderful work being
done here in discovering the Green Man who comes along
with the goddess is also very important. So men are in no
way excluded. In fact, you had Hildegard of Bingen explic-
itly calling Mary a goddess in the twelfth century.
Mechthild of Magdeburg in the thirteenth century calls
Mary a goddess, and she calls our souls goddesses. So the
Mary archetype is in every one of us, men and women.

Then there is the basic archetype of God as mother and
even of Christ as mother. I was shocked when reading
Aquinas to find that even he, who has been pictured as this
great rationalist, calls Christ mother on several occasions.
This is an ancient tradition. Julian of Norwich in England
developed the Motherhood of God theology better than
anyone in Western Christianity. But the whole idea is balance. Eckhart says, "All the images we have for God come
from our understanding of ourselves." That is so important.

If we live in a religious era that does not honor the mother-hood of God and the goddess along with the fatherhood of God, then we have impoverished souls. We are the losers. Because the point of these metaphors is to give us permission to be our fullest selves, to fill that field with something delightful and energizing and re-creating and powerful in terms of healing and creativity.

So a time for recovering the goddess tradition essentially teaches that the creative energies of the Universe and the maker of the Universe are in all things. It's a moment of celebration of creativity. As we know, in that five-thousand-year period in Europe when they worshiped the mother goddess, we have found no evidence of any military artifacts. Instead, what we have found are tens of thousands of statuettes of pregnant women. Of course, when the male archaeologists first saw this they said, Fertility rites and orgies!

But feminist archaeologists tell us that it was a time when the creative process of which childbirth is an archetype was honored. And so to recover that sense of honoring the creative process in all of its forms fits exactly with your theme about creativity, habit, and the evolving laws of the Universe.

Can I take that a little further? In the first creation story in the Book of Genesis, there is the image of God calling forth plants and animals from Earth. God doesn't bring them forth, the Earth brings them forth. Earth is the Mother. Here we have an image of creation through interrelationship. The polarity between heaven and Earth, Father Sky and Mother Earth, is ancient and widespread, and I suppose one of the things we're doing is trying to recover some sense of that creative polarity and also a sense of a source of both father and mother that goes beyond both.

Yes, always remembering we're dealing with metaphors. But the idea that divinity comes from below, not just from above, is clearly part of the goddess tradition and the ancient traditions. So Eckhart says, "God is a great underground river that no one can dam up and no one can stop." It's a very feminist image, that God is not just in the sky. And you have the tradition of the snake: the serpent as the wisdom figure, representing the goddess.

It was so interesting how Hildegard of Bingen dealt with the snake's bad rap in Genesis. She was a Christian and still she had that bad news about the snake. But as a woman who was also in touch with the ancient goddess tradition, she said the reason the snake was chosen to deceive Adam was that no one would believe that this wisest of all creatures would come on and deceive him. So many times in her paintings the frame of her picture is of the snake; she's always redeeming the serpent.

The native peoples of America, the Hopi for example, pray with the rattlesnake. I remember I was once in a kiva with a Hopi praying and I asked him whether the rattlesnake doesn't get very nervous. He said, "Yes, at first, but you sing to it and it picks up the warmth of your song and the warmth of your heart, and then it settles down, and you can pray together for three days and three nights." But he said that because the city was building up out where he lived, rattlesnakes were very rare. It was hard to have good prayer anymore.

The next day I was giving a talk that he attended. He said, "I have to tell you: an hour after you left, a rattlesnake came to my front door. So we're going to pray good prayers in the next three days, and I'm sure it came out of our conversation."

This ancient tradition, which is worldwide, is that the creatures beneath the earth bring special wisdom. It has

influences for our attitude toward our own bodies. The
West has always tried to tell us the spirit is up in our heads
someplace. We want to shoot up through the chakras as fast
as we can to get up into our heads. But the fact is that the
archetype, the metaphors of finding divinity in the Earth, is
also about finding divinity in our own bodies, in our sensu-
ality, in our sexuality, and in the first two chakras that connect to all the vibrations of the fields of the Universe. So it's
a call back to prayer that comes out of our guts and then
reaches our hearts and then also reaches our minds, but it
does not just sit up in our heads.

You talk about the depths of the field and the darkness,
about how we're finding that ninety to ninety-nine percent
of the Universe is dark matter. This whole celebration of the
darkness is part, I think, of recovering a sense of womb
energy, of the darkness of the Earth. Eckhart says, "The
ground of the soul is dark." Whenever we go into the depths
it's dark, whether into the depths of our souls, our fields,
our psyches, or the Universe. We find so much darkness out
there! And there is a word in our tradition that we've ne-
glected for centuries—the Godhead. We're always talking
about God this and God that, but how many theological
tomes have we written or heard, how many sermons, on the
Godhead?

Really there are two faces to divinity. God is the God of
light: "Let there be light," the God of creation and of redemption who works in the light, you might say, and history. But the Godhead is the God of mystery. Eckhart says,
"The God works, the Godhead never works." The Godhead
is pure being. The image I have is a great big cosmic mama
in whose lap all things exist. Eckhart says that our birth is
flowing out of the Godhead, and our death is a return to the
Godhead. When we die, no one will ask us where we have
been or what we've been doing. There is no judgment be-

cause in the Godhead there's such total unity that no one has missed us. Think about that! In the Godhead there is total silence; it is the abyss. In the darkness we're not being judged; we can't be. The analytic brain has to take a nap in the dark. We're being sensed and we're picking up perhaps some of the field energy that we don't catch when the lights are too bright. Or when we have our noses in books.

The Godhead is a male/female metaphor, because God in Latin and in German, which is Eckhart's language, is male. But *Gottheit* and *deitas* are feminine in German and Latin. So I would propose that what the church and our religious traditions have to do is what Eckhart said: "I pray God to rid me of God." And, as he recommended, quit flapping our gums about God. I think we should start moving into God-head language, Godhead metaphor, and Godhead experience, and bring the balance back between the light and the dark, the action and the silence. All this is what the mystics call the *via negativa.*

What do you think of that?

First, the idea of light and darkness is clearly appropriate as a metaphor because the two are related to each other in a polar manner. Light is wavelike, and each wave has a positive peak and a negative trough. (See Figure 5.1 in chapter 5.) The darkness in light is clearly revealed in the phenomenon of diffraction, where light passed through two narrow slits gives rise to a series of light and dark bands. So in light there's darkness in equal measure, oddly enough.

Second, there's another sense in which we think of darkness as connected with death and destruction. Here again there is a polarity. In most traditional religions, the supreme principle cannot simply be a principle of creativity and life, because we have to explain death and decay as well. It seems to me that the Christian conception of God and of

Mary contains both. Mary is not only the life-giving mother, she also, like the archetypal Great Mother, has to do with death. The Pietà, the image of Mary with the dead Jesus, is like this death aspect of the goddess. In fact, the bottom line of the Hail Mary has to do with death: "Pray for us now and at the hour of our death."

So there's a sense in which Mary, especially in the form of the Black Madonna, has some aspects of the dark goddess. In India she is known as Kali. Dark matter in modern physics, when translated into mythic terms, is another aspect of the dark mother. I think materialism in general is an unconscious cult of the Great Mother. Matter and mother come from the same root—in Latin, materia and mater— and throughout the thinking of materialists one can find archetypes of the Great Mother at work: the Great Mother as the economy who can both nurture us and ruin us, in Darwinian evolution as Mother Nature who is both prodigiously fertile and ruthlessly destructive, "nature red in tooth and claw."

And then also we have this image of God as creative and destructive. The biblical God is not all love; he has a terrifying, wrathful aspect. For example, the blessings that come to the Jews at the time of the Exodus from Egypt are the other side of the coin of the curses that fall on the Egyptians. These curses visited on the Egyptians are then mirrored in the dreadful plagues that are to fall on the Earth in the last days, as portrayed in the Book of Revelation.

Almost everyone nowadays is into disaster scenarios. We can all make up our own, using various combinations of overpopulation, AIDS, nuclear threats, pollution, the greenhouse effect, famine, social disintegration, and so on. I wonder how you see all these dark forces that are at work in the world around us.

Perhaps I should go further. I have read you saying that

original blessing came first — there were fifteen billion years of it until people came along — and then original sin happened. But in the Book of Revelation and elsewhere in the Western tradition, and in other traditions worldwide, there is not only darkness on a cosmic scale, but evil. The Fall occurred not only on Earth but in heaven, with the casting out of Satan. In the Apocalypse, the great battle in heaven between Michael and his angels and Satan and the fallen angels gives a cosmic dimension to something that's not just dark in the physical sense of darkness, or dark in the sense of the shadow side, but a force of evil. Just how do you interpret all these things?

First I would like to elaborate on your point about the Mary figure as being vulnerable to suffering (the Mater Dolorosa). The vulnerability of divinity has often been downplayed in the West. We've talked of the "unmoved mover" and so forth, as if divinity is not moved. Rabbi Heschel says that God is the most moved mover, the most vulnerable of all, and so our suffering is the divine suffering. This is a very important dimension to the cosmic Christ tradition: the Christ is not only the light in all things but also *the wounds* in all things. This is Christianity's very great contribution to the world theme of the cosmic Christ. As Jesus says in Matthew 25, when you feed the hungry you feed me; when you clothe the naked you clothe me. So when the rainforest is being killed, we're killing the Christ. That is all about the Mater Dolorosa theme: the *via negativa*, the negative path of the spiritual life as a path of emptying and entering the dark, trusting the dark. But also the path of suffering and letting go and bottoming out. Bede Griffiths said that for many people despair is a yoga. Despair is a spiritual path and they do not come to spiritual experience until they enter the path of despair and disillusionment.

I've seen despair as a spiritual path in a lot of people I've met who have gone into Alcoholics Anonymous. It's through AA that they've discovered a spiritual life, and of course that's a very difficult journey to take. It often becomes a bottoming-out experience, the ultimate letting go that is part of the mystic path. Life begins anew after the sinking and the letting go.

The dimension of cosmic darkness and forces of evil in the apocalyptic literature is very real. Paul says in the Epistle to the Ephesians that our struggle is not against abstractions but against the powers and principalities. Powers and principalities are technical words for angels; these are species of angels in the classical literature. He's saying that we wrestle with angelic forces that are shadow forces. We wrestle with the sins of our fathers, for example. It seems to me we're wrestling with the whole inheritance of industrial mechanization with implications that currently are raining sin all over the planet. What we're wrestling with is bigger than an individual. We can talk about Newton or Descartes or Augustine, but we can't dump blame on any one person—it's a force that's bigger than any of us.

The same is true of racism, the same is true of sexism, of militarism, of homophobia. Aquinas says the greatest sins are not the sins of the flesh but the sins of the spirit. Spirit is often named as angels, good angels and bad angels, in our apocalyptic tradition. There's always a danger in dealing with the apocalyptic imagery of taking it literally, and that must never be done. Satan is not something out there; we're all participating in potential satanic, demonic powers. I would say it's precisely at the level of our creativity that our species makes its choices about whether we will express our divinity through our creativity or the demonic side to our creativity. We are a species that can do tremendous things with our creativity. Whales have been here fifty-six million

years longer than we have and do not find it necessary to
invent nuclear weapons or tear down the rainforest. But we
find it necessary, and we can tear down a huge area of
rainforest in a few days because of our immense creativity.
That's demonic; we're plugging into the creative power of
God but are using it not for life but for destruction.

We have a story of Jesus in Mark's Gospel where he goes
out into the desert, the wilderness, and wrestles with the
demons. It is a vision quest. But, we're told, the wild animals come and the good angels come to succor him and so
together they win the battle against the demons. This too is
apocalyptic in its metaphors. I think it is important to open
up the apocalyptic truths of our time as you, Thomas Berry,
and other scientists are telling us. We are living, really, in
the last times, in the sense that this planet will not survive in
its present form if our generation does not change its ways.
But that is the heart of the Gospel teaching, that we're capable of changing our ways. That is the resurrection I was
talking about earlier: with new images, new science, new
paradigms, deeper prayer, prayer that strengthens our
hearts and therefore makes us bold and courageous so we
can change our ways and let go and begin anew. This is the
very nature of the wrestling with the demons that is part of
any spiritual journey.

*The sin of Satan is essentially the sin of pride, and the fall
of Satan has to do with pride. Is this, as you suggest, a
source of evil that comes into being only with humans? Until you reach a certain level of consciousness, you can't be
proud, at least in any sense we'd recognize. So did the fall of
the angels and the origin of evil in some sense parallel the
development of human consciousness? Are we the source of
evil, sin, and fallen angels?*

Or do we see a demonic element working in the whole of

Nature, present before the origin of humanity? When we look, for example, at some of the more ghastly evolutionary developments of parasitism and disease in the natural world, do we see these as part of a destructive or separative principle inherent in all creation? The primary model we have of creation in the Big Bang theory is of two opposing forces. There's the Big Bang itself, the cosmic explosion that separates everything from a primal unity into outward expansion. This is literally a diabolical force, in that diabolical *means* throwing apart. *Then there is the unifying force of gravitation that is always pulling things.*

This idea of a polarity, of destructive or separative forces and unitive forces, is right there from the beginning in our modern scientific worldview.

I have to question that. Am I not right that that drawing apart of things was a necessary element in the evolution of the Universe as we know it? In other words, this planet would not have happened without that separation happening. So I have trouble calling that diabolical at all. In theological language, what you speak of is what I would call the Eucharistic law of the Universe: everything in this Universe eats and gets eaten in some form or other. For example, the supernova explosion five and a half billion years ago that birthed the elements of our bodies in its explosive and generous death, that was a Eucharistic event. Things die but they pass on their food and nourishment for other moments of evolution.

Therefore it seems to me, using the Christian archetype of the Eucharist, that when divinity passes by, it gets crucified, it also gets eaten but it incorporates. The basic mother image is one of gathering in; it can incorporate even the suffering and the loose ends and the parasites of our world. Isn't it true, for example, that we need parasites in our body to process our food for us? I think that there's a danger of dividing the biological world into good and evil. We can divide it perhaps into comfortable and uncomfortable. If we have a picnic on the side of Mount St. Helens when the volcano goes off, we will have an ideology of the evil of a volcano. But the truth is that a volcano pours special chemicals into the soil making it richer for other generations, and that could not happen any other way.

Mother Earth has a right to let her gas off, as do we. We shouldn't be moralizing about it. There's a danger of moving in too fast with our judgment about good and evil in this regard. And as far as the angels go, they don't live in our sense of time and space. Their fall is certainly a way of talking about our own fall. It's another drama about the depths, the roots, the radicalness of our being here that we can choose to participate in destruction or in life.

CHAPTER TWO

GRACE AND PRAISE

MATTHEW

The issue of grace lies at the heart of an ecological consciousness, for it presents the issue of how we envision our relationship to nature. Grace has often presented a stumbling block for Western theology. St. Augustine in the fourth century not only gave us the term *original sin,* which is not one of my favorite theological categories, but he also split Nature from grace. That dualism, that wound, has haunted Western religion for sixteen hundred years. The creation spirituality tradition has been an effort to bring Nature and grace together.

What happens when we separate Nature from grace? First, we set Nature up for devastation. The ecological crisis has its origin in a theology that separates Nature from grace; if we deny Nature its being graced, then it has no defense when people set out to destroy it for their own ends because it holds no intrinsic value on its own. In addition, when we separate Nature from grace, there is a *grace crisis,* a shortage of grace, a rationing of grace. Anthropocentrism and, with it, institutional bureaucratization set in as people try to ration grace to one another. We are then set up for a clerical system that says: "We are the God-appointed and singular dispensers of grace. Get on your knees and you may get some from us." The sins of the institutional church are clearly encouraged by a theology of grace scarcity, an ideology that separates Nature from grace. In this circumstance religion plays dangerously with temptations to exaggerate its self-importance; it is no longer servant but master; it easily succumbs to an inflated sense of the exclusivity of its place in the world. What follows is a *praise crisis* as well, for when grace is scarce, so too is joy and praise.

I think we can agree that the way we look at Nature and grace is a major issue at this moment of ecological demise and ecclesial crisis. For these reasons I want to share some thoughts with you about grace. Meister Eckhart in the fourteenth century said simply, "Nature is grace." He represents the culmination of the creation mystical tradition that began with Hildegard of Bingen in the twelfth century, then Francis of Assisi in the early thirteenth, then Thomas Aquinas, then Mechthild of Magdeburg, then Eckhart. Eckhart heals what Augustine took apart.

What does grace mean? Grace is about gift. Grace is unconditional love. Grace therefore is about blessing because *blessing* is the theological word for *goodness,* and a gift is presumably something good. A grace is something good, it's a blessing, it's a gift. So when I use the term *original blessing,* you can just as easily use the term *original grace.* The Universe has been a blessing and a grace for fifteen billion years. The amazing drama which has brought this planet to its amazing presence and our species to its presence and all the other species we share the planet with—it all took fifteen billion years of decision making on the part of the Universe—and it was all grace. It was unconditional love. None of us had to prove our right to be here for the original fireball to come up with the right temperature and the right rate of expansion within one-millionth of one-millionth of a second for us to be here. It's what the mystic Julian of Norwich said in the fifteenth century—"We have been loved before the beginning." That's grace—it's unconditional love.

Creation is grace. And we didn't do it. It was someone else setting the table for us.

We have in our language the word *graceful,* and it's a wonderful word. We say, "So and so is graceful," meaning his language is graceful, his movements are graceful, his being is graceful. Notice the *full* aspect to being *graceful.* We don't have a word about half-grace or three-quarters graced. There's a fullness to grace. Indeed, there's a bubbling over to grace. It's like the word *thankful. Grace* and *gratitude* are really the same words. *Gratia* in Latin means *thanks. Gratias agamus tibi* in the Mass—we give you thanks. So gratitude is an intricate part of grace. Meister Eckhart says, "If the only prayer you say in your whole life is thank you, that would suffice."

Being grace-conscious is being grateful. Being graceful is simply living out this basic beauty of the Universe. Gracefulness is a much larger category than is *beauty,* because in our culture to be beautiful you have to be less than so many pounds and you must meet all these other criteria we invent. But some of the most graceful people I've met have been very large people who are graceful in their body's movements, in their dance, in their language, and in their communication. *Graceful* is another word for *beautiful,* but it's a fuller word.

The response to grace is praise. As I said, our civilization is in a praise crisis, we are starving for reasons to praise. Jill Purce, your wife, who has studied chant in Tibet and teaches it to Westerners, says that we have a civilization that doesn't chant. It seems we have nothing to chant about because we've lost the sense of grace and of praise. One reason is that religion is so anthropocentric that it has to keep talking about sin and redemption; it rather weighs one down. Human sin is real, and it is an affront to grace. But that is one more reason we need cosmology—a relation to beings other than ourselves—if we are to feel graced. All the other beings are graced. Recently I went out canoeing with some friends near Cortes Island, British Columbia, and along the way we met four otters on a rock who followed us. We met a woodpecker doing his or her thing on a tree fifty feet from there, and then we met three does. All of it is grace, it's all a gift. Nature is grace.

Someone asked me today: Is the Holy Spirit different from grace? Well, Eckhart has a great line. I don't know what it means but here it is. He says, "The Holy Spirit carries grace on its back." The Spirit graces us. The Spirit might be the source of grace. Grace is the bias of the Spirit in favor of beauty, blessing, goodness. Grace is divine life coursing through history and creation and us.

Onye Onyemici, an African spiritual drummer who works with us in Oakland, says, "When you praise, hold nothing back." When I say we're in a praise crisis I mean we're anal-retentive as a civilization. And there's nothing more anal-retentive than worship. Check it out. Almost everyone in our formal worship in the West is holding back. Our body isn't there, we're holding back; the breath isn't there; the spirit isn't there. Praise is very practical. Getting praise back is about getting our breath back, it's about getting our energy back, and it's about our capacity to let go and to rediscover a capacity for joy and a bigness of soul.

One more point is this: While I like to begin talking about original blessing, original grace, praise, this is not in any way to ignore suffering, evil, or struggle. Just the opposite. It is all to prepare the way for dealing with it. Until we can feel graced—a psychological word for that might be high self-esteem—we're not in a position to deal with our wounds or anyone else's wounds. We run out of steam very early and we give up the struggle. I think the poet Rilke put it wonderfully when he said: "Walk your walk of lament on a path of praise." I'm saying we can't deal with the lamentation, the grief of our time, the grief of the struggle of the Earth, the Earth creatures, the grief of the despair of our young people, the grief of our own wounds caused by racism or colonialism or sexism or homophobia or any other shadow in our collective histories except on a path of praise.

This is why praise is important not only for itself but to get the work with the shadow, the dark work, done as well. Being able to recover the sense of grace provides the context in which we will be able to grace one another—which certainly includes the work of healing and the authentic struggles that we're called to in the 90s.

RUPERT

Grace is difficult to define, but it seems to me to have to do with a sense of connection, openness, blessedness. Openness and connection with what's around us. One evening recently, as I was in an orchard sitting under an apple tree

in the sun, I realized I was there and then in a state of grace.
I started reflecting on the different occasions on which I had
felt this sense of blessedness. Some occasions had to do with
music; others were in beautiful places when I had a sense of
connection with Nature; others came through being in love
and being with people with whom I felt blessed and at ease;
others were through experiences in church services and
rituals. For me some Christian services really are a means of
grace. I've also felt this in Hindu temples and in sacred
places of other traditions. Others came through psychoac-
tive drugs; I have on occasion tried psychedelics, such as
LSD and magic mushrooms, and through them experienced
an amazing sense of beauty. Others have come through
prayer and meditation.

Many different things seem to have been occasions of
grace in my life. It so happens I'm reading Charlene
Spretnak's book States of Grace. In it she shows that at the
heart of many different religious traditions—the Native
American, the goddess traditions, the Christian, the Buddhist—is the mystical experience of grace. I think she makes
a very good case that through states of grace people in all
these traditions have felt a sense of connection with a larger
whole.

It's clear that grace can come in many circumstances and
in different ways. Another form in which it strikes me is
through the Hail Mary, which I say every day: "Hail Mary,
full of grace." This fullness of grace in Mary can actually be
a means of grace, I've found.

Given all this variety, what I'd like to ask you, Matthew,
is how far you think grace is confined to the human realm
and human experience. For example, when birds sing joy-
fully, like blackbirds sing in England, they do so with such
beauty. It doesn't seem plausible that they do it just for us to
hear; they were doing it for millions of years before human

*beings evolved on Earth. And when you look at the beauty
of flowers and the exquisite taste of fruit and wonder of
landscapes, all these things existed long before human beings existed to see them. And I think it would be excessively
anthropocentric to think that these were all there for hundreds of millions of years waiting for us to come along and
appreciate them. It's hardly plausible that they're there only
for the human eye or the human beholder, although we can
share in their beauty.*

One of the more poetic passages in Charles Darwin's On
the Origin of Species *is one where he points out that there
could have been no flower before there was an eye to see it.
Flowers are a response to the vision of animals; they've
evolved in response to animals seeing them. Their perfumes
have evolved along with the animals that smell them. Flowers have been around for at least sixty or seventy million
years, whereas human beings haven't been around for
longer than a million or two. So what do you think about
grace as experienced in the nonhuman realm?*

I have no doubt that grace preceded the human, and
that's what I was saying about the original fireball. The
fireball was a grace. Grace may well outlast humans, too,
especially at the rate humans are going. Animals, too, are a
grace. I chose a dog as my spiritual director for seventeen
years. That's why he was my spiritual director, because he
was truly graced. He never doubted his right to be here and
he was dog. What more can you ask of a dog? He was a
graceful dog in every way. When Merton says every non-
two-legged creature is a saint, he's saying they're full of
grace.

Humans are the only ones who aren't full of grace who
kind of have to work at it. Meister Eckhart preaches on the
line you mentioned from the Hail Mary—"Hail Mary, full

of grace"—and he asks: "What good is it to me if Mary's full of grace and I'm not full of grace?" This is a challenge. I would say that perhaps that's the greatest difference between humans and nonhumans: that the nonhumans are all full of grace and the humans are perhaps like a gas tank that gets full and empty, and we lose grace, we can fall into gracelessness, which may be another word for *sin*. Rabbi Heschel defines *sin* as "the human's refusal to become who we are." We're here to be graced and graceful and instruments of grace to one another. Yet in Bosnia and many other places, we can obviously be instruments of pain and suffering toward not only other human beings but also toward the forests and other creatures.

We're swimming in grace. Examples of this are the flowers you mentioned and the animals who see the flowers. It wasn't just for us, but the fact that we are here and we have the senses to appreciate a lot of this and the mind and the larynx to praise makes us somehow unique. I see our species as a species that was given a larynx to praise with. The other species are busy being graceful, but we are the species to honor all this grace, to praise it. We have to ask ourselves whether we are doing that. Are we good at carrying on that celebration? But your question for me is interesting because that was a question I had for you: Are all beings mystical? It's interesting that in your reflection on your grace experiences you also used the term *mystical experiences.* But do other beings have mystical experiences besides humans?

We know very little about the consciousness of other beings, but most plants and animals spend a lot of their time doing nothing in particular. They're not always rushing around as we are. A lot of people are always rushing around, at least modern people. One of the things that E. F.

Schumacher says in his book Small Is Beautiful *is that the amount of leisure people have seems to be inversely proportional to the number of labor-saving devices. When I lived in India, on visiting villages I found the most wonderful smiles everywhere, even in the poorest places. People seemed to have plenty of time; in the village streets there were people standing and chatting, and no one seemed in a hurry. There is something of this quality in many other rural parts of the world, even in Europe, as in parts of Ireland. But as soon as you get to a big city, everyone's in a hurry, rushing around, filling up their diaries. When everything is speeded up, it doesn't leave much time for grace.*

Some animals seem to be busy a lot of the time—ants for example—but many other animals spend a lot of time just sitting or standing around. I think it's very likely that at least some of the time they feel a sense of connection with the greater whole. But small birds and small mammals like mice are always looking around to see if there's a predator about to pounce, so probably they have a more nervous time. Maybe snuggled up in their burrows they can have a greater sense of connection. It's hard to know. We had a cat and when she was sitting purring contentedly she seemed to radiate a calm, a presence, a joy, which is why my wife, Jill, called her Remedy. She was a remedy for our ills and worries. When birds are singing or soaring, they certainly seem to be in a state of grace. My feeling is that animals probably do have mystical states.

A unique feature of our species is that we can become graceless. When we tear down rainforests, for example, or murder each other in Bosnia, or create inner cities and then ignore the problems there, that's gracelessness. We have the capacity of falling out of grace, and the traditional word for that is *sin.* We are the species that sins. There's no evidence that any other species does. For example, lions eat lamb and enjoy it but they don't go into business setting up lamb stores. They eat what they need; they're not greedy. So we humans are peculiar. Aquinas says that one human can do more evil than all the other species put together. What an affirmation of our divinity and of our demonic side.

We're all born in grace, and the passion for wanting to be graceful at all times is beautiful and wonderful. But again you can't think just about yourself. How about your brothers and sisters, for example, and how are we accomplices in the tearing down of rainforests? If we're still on a meat addiction, hamburgers help tear down rainforests. There are all kinds of connections here. We have to be careful; it's one thing to desire to be in grace and another to be aware that we're capable of opting out of grace, of being graceless or less than graceful. There are degrees. Like the gasoline tank, we can get emptier and emptier. Yet we're never totally out of grace because we're always being graced, we're always being loved by divinity and by the Universe. That is the point: we're never totally empty of grace but we are the species that can prefer our egos over grace, our agendas over the agenda of the Universe.

Grace precedes redemption because grace is coterminal with creation itself. Creation is grace. Nature is grace. Now, redemption is we humans recovering from our acquiescence to evil deeds and therefore you might say to our lowering the quantity of grace in our midst. So it's bringing grace back, it's the return of grace. Traditionally, there is something called the "first creation" and the "second creation." We also need to grow an awareness of being in grace. So awareness is everything. Being awake is everything. It's pretty hard to be joyless in the context of grace. And the reason why our culture is so short on joy is that we're short on grace.

I've never thought of them as synonyms, but I see that you could hardly be joyful without being in a state of grace of some kind. Probably they're closely related.

And also related to praise. What is praise but a response, an utterance, an expression of the joy that's inside that you can't keep down, because joy has to make itself known to the community. Aquinas says that all beauty yearns to be conspicuous. So, all grace yearns to be conspicuous. You taste grace, you want to make it known.

Someone asked me, "What is praise? How do you learn praise, how do you teach praise?" I find that one way to define a spiritual concept is to go to its opposite first. The opposite of praise would be curse or cynicism.

Aquinas says that in a time of cultural crisis—and no one would deny that we're in one today, I think—the rules for morality don't come from a book. They don't come from legal tradition. Ultimately, what you do in a time of crisis to find what's moral is you go to a moral person, you go to a good person. The standard of morality has to be a good person. Then you make up your conscience after such consultation.

Eckhart says, "Who is a good person? A good person is one who praises good people." I read that in Eckhart ten years ago and I've thought about it ever since. It's so interesting, an unusual definition of a good person. What does it mean? To me it means, first, a good person is looking for goodness. That's interesting, you know; everyone isn't looking for goodness. Some people are looking for mistakes, to put people down—operating from cynicism. But a good person is looking for goodness and when a good person sees it, he or she praises. A good person praises good people. The idea is that morality is based on praise, not on good works as such and certainly not on obeying commandments as such. In this spontaneity that praises lies the basis for a rich morality indeed.

Then it is important to recognize good living people.

That would help.

The usual thing is to attack people when they're alive and praise them when they're dead. A lot of the great men and women of history were vilified in their lives and sanctified after they're dead.

Which shows that there weren't that many good people around during their lifetime.

Eckhart's principle also applies in the case of children. Most parents and educators know that if you want to bring out the best in children, you have to encourage them. It's no use just criticizing them all the time. It's a very basic principle that many parents and teachers have grasped because it works.

It always struck me when I was living in Bede Griffiths's ashram how he had time for everyone who visited; he'd invite any visitors along and they'd have some time alone with him. And nearly all people came out of their meetings with him inspired and enlivened. Somehow he had a way of bringing out the best in people. He'd ask them what their hopes and aspirations were, and they'd tell him and he would listen. Then he would encourage the best side of them, with a very positive effect.

To bring that out is the difference between therapy and spiritual direction. There was a study done recently about the practical effects of praise that you mention. The study used a group of young people and was based on the issue of environmental crisis and recycling. Half of them were given only an intellectual basis for understanding the issue. They were taught the rules of recycling, how important it was, all the facts about the effect on the Earth, and even admonished that they had to do this to survive. The other half were just praised and told, "You're doing a wonderful job recycling, you're doing a fine job recycling, keep going, keep up the good work." What they found was that the first group forgot everything; none of them got into recycling. The second group really got into it and stuck with it.

Again, it's the difference between original blessing and original sin. You begin with what's good in people and build it up. Then the goodness increases in its presence. But the other way around just doesn't do the job.

I suppose, though, praise is possible only by contrast with what might otherwise have been. Because otherwise we take things for granted. Most of the time we take for granted the beauty of the sky, the clouds, the stars, trees, the Earth, the air we breathe, and so on. Almost everything's taken for granted until it's threatened. As you have said, having polio at the age of twelve made you stop taking your legs for granted.

Recognizing good people or goodness in nature is done by contrast with what it might otherwise have been. Most of our terms are defined by contrast: blessings and curses, good and bad, and so on. Praising the good involves at least an implicit awareness of the bad. Otherwise there is just a bland neutrality, taking things for granted. In praise there's a sense in which the shadow is in the background.

A part of the opposite of praise you just alluded to subtly is ignoring. In some ways the opposite of love isn't hate—it's no passion at all; it's coldness, it's not caring. Lack of curiosity and wonder—these are sins of omission. This whole attitude of ignoring Nature until it's crashing around us is in some ways a greater insult than coming out and saying, Well, I don't like it, I don't like what trees are doing to me, or something.

I was struck by a passage from Psalm 150. It has a whole new meaning for me as a result of our conversations: "Praise God in our temple on earth; praise Her in the temple in heaven; praise Him with blasts of the trumpet; praise Him with lyre and harp; praise Her with drums and dancing; praise Her with strings and flutes; praise Him with clashing cymbals; praise Him with clanging cymbals; let everything that breathes praise Yahweh."

The idea is that praise is not just anthropocentric, it's us joining the cosmological praise that's going on. Hildegard says that every being is praising God. A beautiful passage from her writing says, "The fire has its flame and praises God; the wind blows a flame and praises God; in the voice we hear the word which praises God; the word when heard praises God, so all of creation is a song of praise to God."

Praise is going on all the time; humans have just opted out most of the time. We're just bitching, cutting each other up and being cynical, wallowing in our self-pity. Self-pity banishes praise, doesn't it?

It does. But how do you think it's possible to turn this around? How can we become more aware of the beauty in Nature, the grace in Nature, and the praise in ourselves?

Science is absolutely essential to tell us what's praiseworthy in Nature. To hear that Gaia has arranged things, that even though the sun has increased twenty percent in its heat, our planet has maintained its comfortable temperature. Clearly there is some kind of air-conditioning going on, as James Lovelock points out in his Gaia theory. That's worthy of praise since without that we wouldn't be here to think even bad thoughts, never mind praiseworthy ones.

And that's where the story of Creation is absolutely essential. It takes a scientist to teach us to praise Nature because the scientist has the word on the wonders of Nature.

I'd rather say a naturalist, because naturalists are those who look at the phenomena as we see them: the stars in the sky, the different kinds of plants and animals around us, the migrations of birds, and so on. This is the kind of science that can most directly lead to praise.

It includes the heart, doesn't it?

That's right, and it's no coincidence that in nineteenth-century Britain many of the amateur naturalists were country vicars. Naturalism and religion were considered to fit together very well, as indeed I think they do. There's a very strong strain in science, though, which removes praise. The neo-Darwinian theory of evolution, for example, says that everything you see, however wonderful it may seem, is merely a matter of chance mutation and natural selection, the purposeless product of blind mechanisms.

It stuffs the praise back into your body, doesn't it? Even if you're feeling grace, you don't dare let anyone know.

And the feeling you get from it, the take-home message, is not how wonderful Creation is, but how smart we are to have figured all this out.

So we praise our geniuses, just a few human beings, and that's it. Then we die.

Yes. We're praising only a select few, like Darwin, and they're mostly men. Instead of it all being for the greater glory of God or seeing the grace in Nature, we're actually saying how smart scientists are.

We also praise military heroes, too, who win wars for us. We put their statues very large, looming in our cathedrals even. I remember that the first time I visited St. Paul's Cathedral in London, I was shocked. I went up this aisle and here's some soldier who killed thousands of Indians in a battle in the Punjab and here's one who killed thousands in another battle. Right in the heart of a cathedral.

It doubles up as a sort of national Hall of Fame. But while we're talking about praising people, here's a related idea that came up recently. I was riding on Lake George, New York, in a boat with some friends. One side of the lake is stupendous, just forest and mountains. The person we were with told us that the whole area is forever wild, it's been preserved. So has a lovely island in the middle of the lake, an old sacred place for the Native Americans. I thought, thank God these have been preserved, rather than developed as powerboat marinas and that kind of thing.

Then the idea came, what about having in each country a national conservationist day? Just as we honor the war dead, we should honor those who have done so much to save the natural environment—John Muir, for example. Each local community would have a roll of honor of those who have preserved parts of the local environment.

I and my family live next to Hampstead Heath in London, and whenever I walk on it I feel grateful to the people

who fought to preserve it over a hundred years ago. Other-
wise it would have been gobbled up as London expanded
and developed. All of us who live near there are perennially
in their debt. But at present we don't acknowledge them.
On the national conservationist day we would.

Another thing would be a ritual like John Seed and
Joanna Macy have developed called the Council of All Beings, which takes you into grief about the disappearance of
species. But why not a ritual that praises species where, for
example, humans sit around in circles and make masks and
become giraffes and become bears and become oak trees
and so forth? Not in terms of grief—which is important—
but in terms of praise. So we can praise the giraffe and
praise the eagle and so forth. That would be a very simple
thing to create a liturgy around. The word *Eucharist* just
means *thank you;* it is the word for thanks or praise in
Greek.

As to the original question, how do you turn liturgical
language into something real, we've got to get away from
words alone and start doing things. All children love plays.
We should start playing at these important issues like teaching praise. Then bring out the wonderful hymn from the
psalm. Instead of talking about lyres and cymbals, bring out
the music, the bands, to cheer on the giraffes and so forth.
There you'd have a liturgy that someone would come to.
Maybe even the giraffes. In the Cathedral of St. John the
Divine in New York, every year on St. Francis's feast day
they invite the local circus in. They have snakes, lions,
bears, and everything. It's the blessing of the animals. They
even have elephants come into the cathedral, followed by
someone with a shovel of course.

Yet I've always been repelled by the notion that animals
need our blessing. Animals don't need our blessing, we need

their blessing. It should be blessing *from* the animals—that would be the real praise, instead of blessing *of* the animals.

You get that in Hindu temples. The great Hindu temples in southern India usually have a temple elephant. You can stand in front of it and this huge animal then puts its trunk on your head to give you a blessing. Then, because Indians are very practical, the trunk comes around in front of you and waits until you put a coin in it, which the elephant then gives to its keeper. This helps to pay for its food, and it does eat a lot.

I've heard of a monkey asking for money, but an elephant—that's got to be irresistible, and beautiful! Contrast that to a zoo where you're gawking; here you're being blessed by and honoring the elephant.

The elephant is the living presence of one of the great Hindu gods, Ganesh, who removes obstacles.

What do you think Western cathedrals would be like if they had an elephant? If cosmology was truly represented not just on the walls and so on but in the flesh?

I think they'd have to be indigenous rather than exotic animals, such as horses, which were sacred animals in pre-Christian England. Most English people are still quite horrified by the idea of eating horses.

It's an interesting fact that among domesticated animals we've got two categories. One are seen as a means of grace and we certainly wouldn't eat them—namely, dogs, cats, horses, and other pets that are treated like honorary people. Billions of dollars a year are spent on their well-being— veterinary care, pet foods, toys, and even clothing. The other category of domesticated animals includes sheep, cows, pigs, and chickens, which in the West are treated with no respect whatever. They're units of production in factory farms. One category consumes pet food; the other category is processed into it.

A small part of the animal kingdom is already accepted in many people's lives as a means of grace. People see and enjoy the grace of cats as they move and as they purr and as they provide wonderful comfort to many people. People obviously love the energy and companionship of dogs as well.

Probably for most people in the modern urban world, the main teachers about the animal and plant kingdoms—their main experience of natural grace—are their pets and their gardens. In England especially, gardening is far and away the most popular hobby. For many people a communion with the plant realm is a matter of daily practice in their garden.

A place of angels, the angels in the gardens.

Gardens and pets are very important channels of communication to the nonhuman realm, a much-valued part of many people's private lives. But they find it hard to acknowledge this in public.

You would think that this would be one of the church's roles, to make more public and somehow introduce these realities into the sanctuary, that part of the tabernacle.

Well, in our parish church I suggested to our vicar that we have something like this and he agreed and started doing it on the Sunday after St. Francis of Assisi's day. The form he chose was very good—including praise and thanks for the pets and all the gifts they give us. He really loves his

dog, so this came from the heart. Then we had prayers for
all the animals that were being maltreated and suffering in
the world, calling them to mind. All the pets present were
then brought up to the sanctuary and he blessed them, and
their owners seemed very happy for this to happen.

Did any fights break out between these animals in the
sanctuary?

No. There were just a few barks from the dogs.

But the cats didn't chase the dogs and vice versa. Isn't
that interesting that the animals found some peace in
church. I remember they did that at an Episcopal church in
downtown Chicago ten years ago when I lived there and
they never had so full a church as they had on that day.
Everyone loves to bring their pet, especially people living in
the city. And a church will get filled not only with animals
but with humans.

Pets link people intimately to the nonhuman realm. But at
the same time there's also a taboo against pets. This was
already evident centuries ago, as in the denunciation of
witches' relationships with their familiars. It still exists to-
day in the widespread, largely unconscious feeling that to
lavish attention and affection on pets is to waste what
should properly be lavished on other human beings. Rela-
tionships with pets are regarded as an inferior substitute for
proper human relationships. This attitude leads to a sense of
guilt about pets, which is a major reason why people's rela-
tionships with their animals are kept in the private realm.

Maybe that comes from the idea that we have only a
limited amount of praise to lavish or give. That is silly, of

course, since love increases love; it's not a material kind of thing that has only a certain amount and then it's gone.

My sense is that it arises partly from our carnivorous habits and partly from the domestication of animals. Ever since the domestication of animals some ten thousand years ago, people have exploited animals. Horses were forced to pull carts, oxen were forced to draw plows, and if they didn't, they'd be whipped. Many domesticated animals were eaten. Consequently, if you acknowledge any kind of community or relationship with animals, then it's difficult to exploit them or to eat them.

People in shamanic hunting cultures have a different relationship with animals. The animals have spirits they can acknowledge and praise and are the focus of rituals. When these people hunt the animals and kill them, it is a sanctified act in some sense.

In traditional religions, the killing of the animals is also a kind of sacrifice. For example, Muslims have a festival called Bakr' Id, commemorating Abraham's sacrifice of the ram instead of Isaac. Each family gets a goat or a sheep, and then the head of the household kills it ritually with everyone looking on; then they eat it. The killing of the goat and the shedding of its blood are literally brought home as a reality. This is a sacrifice. Whereas modern Western people just go and buy meat in supermarkets, neatly packaged in plastic wrap. The killing goes on behind the scenes, and I dread to think what it's like.

In the forefront it's cleaned up. Jeremy Rifkin in his book on the beef industry points out that our culture reduced cattle to an industry and also developed our beef addictions when we took the sacred element away from the bull. We

secularized cattle, and the result was an unhealthy and avaricious industry.

The shamanistic attitude toward animals you mentioned
connects to our notions of sacrifice. Perhaps one reason
we've lost the meaning of sacrifice is that during the last
three hundred years we anthropocentrized sacrifice and
then dispensed with it. Except for wars, wherein we do
indeed offer sacrifice: we sacrifice our young people in wars
quite readily. But maybe the clue to authentic sacrifice is
realizing the cosmological event. Every time we eat food
there is sacrifice. That's why I talk about the Eucharistic
law of the Universe, that we are always eating and being
eaten and so is the Christ, so is the divine presence in this
space and time called the Universe. There is a law here of
sacrifice. Even divinity gets sacrificed, and we've cut ourselves off from that.

CHAPTER THREE

THE SOUL

RUPERT

The soul is the animating principle, that which makes living things alive. In Greek it's called the psyche. *We now think of the psyche as the human mind, but for the Greeks the psyche had a far wider meaning: it was the life principle of all living things, including plants. The Latin word for soul is* anima, *the source of our word* animal. *We talk about* animate *things as opposed to* inanimate *things: things with souls as opposed to things without souls. The traditional meaning of the word* soul *is far wider than the human soul. The soul is that which makes things alive.*

A starting point for any reflection on the nature of life is death, comparing the dead body of a person or animal or plant with the living state that preceded it. The amount of matter in the dead body is the same as in the living body, the form of the body is the same, and the chemicals in it are the same, at least immediately after death. But something has changed. The most obvious conclusion is that something has left the body; and since there's little or no change in weight, that which has left is essentially immaterial.

Similar thoughts must have crossed the minds of people in different cultures because very similar concepts are present everywhere. Often the soul is related to the breath. And the breath isn't just the air, it's the whole process of breathing in and out, the animating activity.

In the animistic traditions of the world—which means all traditions except the modern West for the last 350 years—it was taken for granted that many things in nature are alive besides ourselves. All plants and all animals, the entire Universe, the planet Earth, the other planets, the sun, the stars. All of these were thought to be living beings, all with their own kind of soul. This was the tradition that the Greeks inherited and formulated in philosophical terms. The Platonists spoke in terms of the anima mundi, the soul of the world, the whole cosmos being a living being with a body, soul, and spirit. Aristotle formulated it in biological terms. He spoke about the souls of plants as governing the forms of the plants as they grow. The soul is the formative principle. According to Aristotle, an acorn grows into an oak tree because it's pulled toward the final form, the mature form of the tree, by its soul. The soul contains the goal or what could now be called the attractor of the developmental process. The soul has the end (Greek: telos) in itself; and Aristotle also called it the entelechy.

In the case of animals, the soul plays a formative role, just as it does in plants. In English this aspect of the soul was called the vegetative soul. As the animal embryo grows, it gives the form to the body, and it also underlies the maintenance of the form of the body, the healing of wounds, the *regeneration of organs after damage, and so on. In animals, there is also the animal soul, which is concerned with the correlation of the senses and the movements; it acts as the coordinating principle of the instincts and of behavior in general.*

In human beings, there are three aspects or levels of the soul: the vegetative soul, responsible for the form of the body; the animal soul, giving us our animal nature; and the intellect or the rational soul, the conscious part of our mind, our reason. In classical thinking, the conscious mind was part of a much larger psychic system which linked us to the animals and plants.

A great deal of this traditional understanding was incorporated into Western Christian philosophy by Thomas Aquinas and others and became the basis for the medieval doctrines of the nature of the soul.

A lot of the present confusion began in the philosophy of René Descartes in the seventeenth century. Descartes said that the whole of Nature is a machine; it is inanimate. In effect, he withdrew the soul from Nature, from all animals and plants, and from the human body as well. Before that, the soul was believed to permeate the whole body. And the body had different psychic centers within it. These are the chakras in Eastern philosophies. But in the West there was a very similar understanding. We still have it in Christian liturgies that place the heart at the center of the deepest thoughts. In the Anglican prayer book, the Collect for Purity starts, "Almighty God, unto whom all hearts are open, all desires known and from whom no secrets are hid, cleanse the thoughts of our hearts by the inspiration of thy Holy Spirit." In the Magnificat: "He hath scattered the proud in the imagination of their hearts." The heart was a psychic center and a center of the deepest thoughts, as it still is in

Tibetan Buddhism, for example. But for Descartes and his followers, it was merely a pump, part of the mechanism of the body.

What happened as a result of this Cartesian revolution was that the psyche, the soul, was withdrawn from the body, and the body became a mere mechanism. The vegetative and animal aspects of the soul were abolished. All that remained was the rational soul that made humans distinct from all other creatures. And this was confined to a small region of the brain, the pineal gland.

How the rational soul interacted with the machinery of the body was an utter mystery to Descartes and it's still a mystery today, because today we still have the same theory. Cartesian dualism has become the standard theory of our culture. The only difference is that the supposed seat of the soul has shifted a couple of inches from the pineal gland into the cerebral cortex. But we're still left with an animating principle inside the brain somehow controlling the body. This is sometimes called the "ghost in the machine" model of consciousness. Materialist philosophers, like Daniel Dennett, for example, try to exorcise the ghost from the machinery. But they find it very hard to convince people that they themselves are computerized automata.

In spite of all the efforts of materialist philosophy, Cartesian dualism remains predominant. In the Natural History Museum in London, in the section on human biology, there is an exhibit called "How You Control Your Actions," with a model of a man with a window in his forehead, through which you can look. Inside the head is a complex array of dials, controls, and levers—the cockpit of a jet airplane. And then there are two empty seats, presumably for you, the pilot, and your copilot in the right hemisphere. This says it all. I've been there when parties of schoolchildren are being *led through so that they can understand what contemporary science has to teach us about human nature.*

I'm sad to say, my five-year-old son's kindergarten classroom has a standard schoolbook called How Your Body Works. And there are pictures and diagrams of the body full of conveyor belts, escalators, pumps, tubes, and so on. Inside the brain, there are lots of little men pushing plugs into telephone exchanges, looking at dials on computers, pressing buttons, and so on.

This is the usual modern view, and of course it's absolutely ridiculous. The so-called mind/body problem is so notoriously intractable that most scientists think: Why bother with it? The result is that you can be educated as a biologist, as I was, and you can teach in a medical faculty, as I did, without ever mentioning the word consciousness. It doesn't come up. If somebody raises it, the usual response is: "That's a philosophical problem best left to the philosophers. They've been talking about it for centuries and they've got nowhere, so we won't waste time with that. It's more productive to get on with the job of finding out more about the mechanisms of the brain."

When Descartes made this division between the realms of matter and spirit, he established a new demarcation between science and religion, defining their territory. Science took the whole of nature, including the human body. All nature was secularized. The arts and religion took the soul, this ineffable something located in a small region of the head. In this way a modus vivendi was established between science and religion. Science was concerned with the objective realm of facts, religion and the arts with the subjective realm of values, aesthetics, morality, and belief.

Science got the better part of the bargain, since it got practically everything, as defined in its own terms. The soul was confined to a small region of the brain, and materialists regarded it as nothing but an aspect of the functioning of the brain, inevitably coming to an end with the death of the body and the breakdown of the cerebral machinery.

This is the legacy of Cartesian dualism. But I want to end by saying that I think we may be able to go beyond it. To understand how this may be possible, we need to recall the older, pre-Cartesian understanding of the soul. The old view of the soul was very general. It didn't apply just to animals and plants, it applied even to magnets. Thales, the first of the Greek philosophers, said the lodestone has a soul, the magnet has a soul.

A magnet can act at a distance, attracting or repelling things, without anything in between. And if you heat a magnet, it loses its force. It's as if there's an animating principle in a magnet that can disappear, just as the animating principle in an animal or plant can disappear. So in the ancient world, and right into the seventeenth century, it was generally believed that the magnet had a soul. The basis of magnetism and electricity was a psychic reality, a psychic entity. William Gilbert, the founder of the modern science of magnetism, whose great book on the subject was published in 1600, still thought in terms of magnetic souls; the compass needle pointed to the north because of the magnetic soul of the Earth. The soul was part of the common discourse of science right until the time of Descartes, when souls were banished from all of Nature.

But magnetism remained. Like Newton's gravitation, it worked at a distance. But how? It was a profound mystery that no one could explain. They just forgot about the problem. It wasn't until the nineteenth century that Faraday replaced the old idea of the soul with the new idea of the field. In the 1920s, Einstein extended the field concept to gravitation; the gravitational field coordinates and holds together *the whole Universe, fulfilling the role of the anima mundi, the soul of the world, in the old cosmology. Then fields were introduced in the heart of matter, as quantum matter fields. Fields were introduced into biology in the 1920s in the form of morphogenetic fields, form-shaping fields, invisible formative fields organizing the development of animals and plants, maintaining them and underlying their regeneration.*

According to the British biologist C. H. Waddington, morphogenetic fields have attractors within them. So the oak tree has a morphogenetic field containing the attractor, Aristotle's entelechy, the mature oak drawing the developing acorn toward it. In effect, the vegetative soul was reinvented as the morphogenetic field. Everything the soul did before, the field did now.

So souls have been brought back into science in the guise of fields. My own work is concerned with morphic fields. The morphic fields of organisms underlie not only their form, but also their behavior; they play the role of the animal soul. They also underlie the activity of the rational mind.

MATTHEW

Thank you for your lucid walk through the pathology of Western philosophy. The story is so sad, because we've said not only that the trees out there aren't alive and the fishes aren't alive, but also that they don't have souls, the animals don't have souls, we're the only ones left with souls. No wonder we're so lonely! In the process, we die of the cosmic loneliness that we've created for ourselves in our anthropocentric arrogance, which is a contemporary way of describing the spiritual sin of pride. That's a real question for humans. You raise it in terms of life after death, but if we come to some authentic experience of soul in this lifetime, how could it cease? Hildegard of Bingen says, "No warmth is lost in the Universe." I would say that no beauty is lost in the Universe: that's what the Communion of Saints is, a gathering of beauty. It's not just two-legged saints; in fact, it's all the saints.

Your point that the Cartesian model is still prevalent is very important. It's running all of our academic institutes, including seminaries. In that model, God is the engineer or the mathematician outside Creation. This is theism: God is way out there someplace, as far away as he can get. Contrast that to the image of God from Aquinas, who says, "God is life *per se* life." That's the whole animistic tradition that you said reigned right up to the seventeenth century. If God is life *per se* life, God is in the midst of everything that's bubbling, everything that's vital. And that is panentheism, to see God in all things.

You made the point that the heart, the human soul, was pulled out of the human body by Descartes. In contrast, we have a tradition of chakras in both East and West, with an emphasis on heart. I would like to point out a sentence from Aquinas that is startling. He says, "The objects of the heart are truth and justice." Descartes said that truth is found only in clear and distinct ideas. It's way up there, it's not even in the pineal gland, it's next to the pineal gland. But for Aquinas, representing the more ancient tradition, the truth is a passion of your heart. It is a heart issue. And so is the quest for justice.

I really love your writing about the fields and the soul as field. This is a profound contribution. Because the question of what soul means is a critical one. We have lost the meaning of soul in the West. It's been trivialized: it's the pineal gland, then the cerebral cortex. It's just so puny, who really cares? Charles Fair is a philosopher of history who says that when a civilization loses its meaning of soul, it breaks apart as a civilization. Soul is a most important concept. It gives meaning to the essence of civilization. When a civilization loses its meaning of soul, it's the end of an era. The good news is that we're at the beginning of a new era if we can come up with some new images for soul.

We're at a whole new level today of finding what soul means. Your image of fields is so inviting. What is a field? It is a space for playing in. A field is a place for running in, frolicking in, dreaming in, soaring in, stumbling in, investing our passions in. I think all of those are images of soul. To me, soul is our passions. It's what we love. As Jesus said, it's where your treasure is. Our treasures are in the fields. Maybe they're hidden, the field is not all aboveground, and that's part of the invitation to lie down in the field and play in the field. Our souls are where we play and of course also where we suffer, where we fall down, where we fail. The potential of this image of field is untapped. You rightly put it in a nonanthropocentric context. Not only does every individual, plant and animal, have soul but mountains and places have soul, and communities have soul, groups create soul. The biosystem itself and Gaia and our Milky Way and cosmos have soul since they have togetherness.

For me, some of the questions of soul for humans would be these: How do we awaken the soul? How do we heal the soul? How do we guide, that is educate, the soul? What if education were about guiding the soul? And perhaps most important of all at this time in history, how do we enlarge the human soul? Eckhart says that God is delighted when your soul grows bigger and bigger, which is another way of saying that it's time to grow soul.

We have trivialized soul not only in the physiological and mechanistic sense, but in every other sense too. Our souls shriveled up during the Cartesian era. If you're cut off from the souls of all the other animals and plants and beings and stars, you're just hiding away inside your own little man-made space. It's not even a space, it's a place. No wonder we have *acedia* (boredom), that sin of the spirit that comes, as Aquinas says, from the shrinking of the soul, the shrinking of the mind. We have shrunk our souls. That's the process of the last three centuries. That's why we can tear down rainforests, destroy our nest, kill the children to come and not think of it, and can say we're all holy and happy and full of bliss. We're out of touch with our own grief! We are separated from the responsibility for our own actions and our connection to all the joy and pain of the world. All this is about souls that have shrunk up. This is why we're into addiction—we're covering up the truth of our shrunken souls with everything from liquor to drugs to television to shopping to sex to anything.

Our souls are too small. We suffer from what Aquinas and the others call pusillanimity, a puny soul. Whereas we're all here for magnanimity. We're here to become a great soul, *magna anima.* Pusillanimity commits itself to fear and to contraction. That's what fundamentalism is all about—it's about contracted souls. Anthropocentrism is about contraction, a refusal to connect to others, to the greater Universe, to cosmology.

Now, how is soul related to the body? I just want to say this one sentence. I have found it in Hildegard from the twelfth-century, Aquinas from the thirteenth century, and Eckhart from the fourteenth century. All of them have this sentence: *The soul is not in the body but the body is in the soul.* It's a very important shift of consciousness, and it helps correct Plato's mistake of the soul being a bird stuck in a cage with the result that we're not really going to get free until we die. That whole notion is quite dangerous. Our body is in our souls means that our souls are as large as the world in which we live, as the fields in which our minds play, as the fields in which our hearts roam. That's how big our souls are.

Now take the fundamentalists: they have a square box. That is the world in which they live. The world you live in is the world you believe you live in. So the fundamentalists' cosmos is quite small, and it shows. The cosmos we're being invited to play in today is one trillion galaxies, each with at least one billion stars. Each with a fifteen-billion-year history. How can we ever be bored again? We are invited to roam in this huge field today.

If our body is in our soul, then our body is an essential instrument that celebrates and praises, expressing the soul's passions: delight, wonder, joy, desire, grief, pain, suffering. The body is an instrument for our soul, and that's why when the body is missing from worship it's so boring. Aquinas says that we sin with the soul, not with the body. It's the *choices* we make; there's nothing evil about the body. He also says that virtues such as magnanimity are in our passions, they're not in our will. Magnanimity is not about willpower—I'm gonna be a big soul today—it's about following our passions and guiding them to the directions that we're called to.

What about soul and spirit? It's very important to point out that soul and spirit are not the same. Spirit is in matter as much as in soul. Dualism is based on this notion that soul and spirit are the same thing. St. Augustine said: "Spirit is whatever is not matter." He takes spirit entirely out of everything that's matter. Aquinas says that spirit is the vitality in every being, in all matter, including in our souls. So spirit is in soul, we hope. But not if our soul is dead or barely purring. That's the point. Body and soul are ours, but spirit is not. We are body and soul, but spirit is greater than our bodies and souls. It blows where it wills (John 3:8). We're not in control of spirit. Spirit is everyone's and no one's. No one possesses spirit, it's not private property. No church owns it; no religion owns it. It's greater than the soul. Spirit requires receptivity, an open heart, a letting go.

Eckhart says that the soul grows by subtraction, not by addition. One way the soul grows is by grief, by suffering. Other ways the soul grows are by praise and by meditation and by simplifying our lives and being with being. Spirit is a wild river, a raging fire, a great eagle, a great goose. Spirit is something very wild, and we can get ourselves prepared for it, to receive it.

What about the relationship between soul and heart? I think that of all our words in the West, *heart* comes closest to what we mean by *soul.* As you said, Rupert, soul is not just psyche or even mind. The word *mind* in the Middle Ages was about creativity. Rationality included our creativity at that time. When our hearts break, our souls break, our whole cosmos is shattered. The word *heart* is about as close as we have to the word *soul,* at this time.

The mystics are artists of the soul. They are poets of the soul. That is why anyone who is serious today about soul work, including therapists who are getting serious about it again, are drawn to the mystics.

My friend Meister Eckhart has this incredible sermon on the soul. It may have been his very last sermon because it's so profound and mature. He says: "A master who spoke the best about the soul says that all human science can never fathom what the soul is in its ground. To know what the soul is you would need supernatural knowledge." In other words, the soul is ineffable. It's so deep you can't fathom it, it's bottomless.

Eckhart continues: "We do not know about what the powers of the soul do when they go out to do their work.

We know a little bit but not very much. What the soul is in its ground no one knows." He is celebrating the essential mystery of your soul. Eckhart says further, "What one can know about the soul must be supernatural. It must be from grace, *for the soul is where God works compassion.*" He says that we don't have soul until we become a field where God is working compassion. Or, to put it differently, we're not born with a soul, we have to make it, in a sense. In a sense, you make soul through living, living joy and living grief, and out of both compassion is born.

I think Eckhart's statement is one of the most amazing statements in Western history: the soul is where God works compassion. Think about it! We don't have soul until we become a field through which compassion is working. This is not just Eckhart. This is the richest mystical tradition East and West, that soul is compassion, the work of compassion. Compassion is work, it's not just being there.

Teresa of Ávila's finest book, I think, is *Interior Castle*, her book on the mansions of the soul and the rooms in the mansion of the soul. She goes through seven rooms but she ends her book this way: "Now we've explored seven rooms in your soul, but in fact your soul has millions of rooms, most of which never have their doors opened. And in every one of them there are labyrinths and fountains and jewels and gems and gardens." I just love that. I remember once I was lying on my living room floor listening to opera and in the middle I said, Why am I listening to opera, who needs opera? Then that line from Teresa of Ávila came through— we have to open up these doors of the soul, and maybe even opera can do this. Why die before we've opened at least a million doors! The fourth room is where we learn silence and we learn to let go in the name of God and quit praying with words. She says that most people never get to this fourth room: they still think prayer is talking to God and having someone talk to God for them. Until we can get into our silent heart we're not going to enter the fourth room.

Rumi, the late-thirteenth-century Sufi mystic, practically a contemporary of Eckhart, also has a lot to say about soul. Here's a good example from Rumi about silence. He says: "Secretly we spoke, that wise one and me. I said, 'Tell me the secrets of the world.' He said, 'Ssssh, let silence tell you the secrets of the world.'"

One point about soul and exploring the soul is that it's an inner journey. This needs visualization. Eckhart says that in the innermost depths of your soul God creates a whole Universe. All past, all present, and all future come together in the innermost depths of your soul. The soul experience is about going innermost, going to the center. Rumi says: "There is a forest within, which gives you life. Seek that." (That's exactly what you are saying: soul is the animating principle.) "In your body lies a priceless gem, seek that. If you want to find the greatest treasure, don't look outside, look inside and seek that." But remember: inside also means looking into your dreams of the fields and into the Universe. Look into wherever your soul lives, which can be very far. But the point is, don't be superficial; don't be on the outside, be on the inside. That is Eckhart's model, which he gets from Paul. It is all about the relationship between the inner person and the outer person or, in today's psychology, the true self versus the outer self.

Finally, Rumi says: "All my talk was madness filled with do's and don'ts." Morality is an outside thing, it's do's and don'ts, commandments which get boring after a while because we can memorize them in ten minutes. "For ages I knocked on a door. When it opened I found I was knocking from the inside." Is that any different from Jesus saying, The kingdom and queendom of God is among you. We've been knocking from the inside. Or, as Jesus was saying, knock and the door will be opened. What door is it?

In the modern world, when people talk about the inner life, they mean a life somewhere inside their body and especially inside their brain. Then there's the outer life, which is the whole of the external realm. We've internalized this shrunken soul and think that the inner means something inside our brains and the outer is something out there. Well, if the soul is extended or expanded, as you said, the inner isn't confined to our bodies; it's the inner experience of things. Science is based on trying to do everything with the outer experience of things. The objective method is supposed to remove all psychic interests of the scientists from what they're doing. In fact, most scientists are heavily engaged in what they're doing if only for reasons of personal ambition. The idea that they're totally objective, like disembodied minds, is a pretense.

This idea of the inner involvement being not just inside but outside us is the key to breaking down the Cartesian division. What you call the shrunken soul I call the contracted mind, the idea that the mind is contracted into the brain, as opposed to the older view which saw our souls as extended all around us. Our soul was not confined to our head. It not only permeated the whole body but was involved in all experience and perception.

I believe that the extended mind can actually be approached empirically. In my book Seven Experiments That Could Change the World, *I suggest some simple experiments that can be done very inexpensively but that have profound implications for modern science. One of them helps to show we're not just playing with words here; we're talking about things that can actually be tested scientifically.*

The normal view of perception is that when you see me,

light rays come from me, go into your eye, an inverted image forms on your retina, electrochemical activities happen
there, and nerve impulses go to the brain, where complex
patterns of electrical and chemical activity take place in the
cerebral cortex and elsewhere. All this can be observed and
measured empirically. Then, somehow, in a totally unex-
plained way you form a subjective image of me inside your
brain.

The image forms inside your brain and then, even more
mysteriously, you imagine, falsely, that the image is not inside your brain but here, outside you, where I am sitting.
That's the standard view of perception. To accept it, you
have to deny your most immediate experience. You don't
experience this room and me sitting in it as being inside
your brain; you experience it as being outside your body.

What I'm suggesting is a hypothesis so devastatingly simple that it's difficult to grasp. I'm suggesting that your image
of me is just where it seems to be—outside you. It is indeed
a mental image, constructed and interpreted by your mind.
But it's not located inside your head. As well as light rays
coming in, something's going out: your perceptual world.
Projection is perhaps too crude a term, but it's the easiest
one to use. We're projecting out images as well as taking
them in.

When we look at somebody or something, there's an outward projection of images which usually coincides with
where the person is—your projection of images on to me
coincides with my actual position. If it were not in coincidence, then it would be an illusion or hallucination. Again, if
this is not a mere play with words, if there is an outward
movement of the mind to touch that which is perceived,
then perhaps we can affect things or people just by looking
at them.

Can we test this? Can we affect people just by looking at

them? For example, if we stare at somebody from behind when he can't see us staring at him, can he feel we're looking at him? Is there such a thing as the sense of being stared at? The answer of course is yes. It's a well-known phenomenon. Most people have experienced it directly.

When I first thought of this and its implications, I looked through the archives of psychology and parapsychology for research on the feeling of being looked at. I found that practically no research had been done on it, even by parapsychologists. It's been extraordinarily ignored. The reason is that it's so uncomfortably close to the folk belief found all around the world that you can affect people by looking at them. If you look at somebody or something with anger or envy, you can blight what you look at. It's called the evil eye. In modern Greece, for example, you see people going around with blue eye-shaped amulets. They are direct descendents of ancient Egyptian talismans of the eye of Horus, the sun god. The eye of Horus even appears on every dollar bill, on the Great Seal of the United States: the radiant eye within the triangle at the top of the pyramid.

The converse belief, that looks can confer blessings, is also widespread. In India, for example, it underlies the tradition of darshan. You go and see a holy man or woman for his or her look, darshan. Just being in that person's presence and having him or her look at you is believed to confer great blessings.

I've devised a very simple experiment to test whether or not people can tell when they are being looked at. You look at someone from behind, or you don't, according to a random schedule. The person has to guess whether or not she's being looked at. This test has now been done thousands of times and the results are clear. The effect really does seem to exist, implying that our psychic world, our mental world, isn't just inside us, it extends around us.

Jean Piaget, the Swiss psychologist, studied the way chil-
dren's ideas develop, and in his book The Child's Conception of the World *he shows that until the age of about ten or*
eleven children think that their thoughts are in their breath,
in their throats, or all around them. But they then learn
what he calls the "correct view," namely that thoughts are
invisible things inside the head. Since no one has ever seen a
thought inside a head, this is an amazing act of faith that
virtually our entire civilization has come to take for granted.

By contrast, if we admit that our souls are extended all
around us, what about the souls of dogs, squirrels, deer,
birds, and other animals? They're also looking around them,
and their mental fields must be extended outward into their
environment. We live in a world of overlapping mental
fields, of a shared space which is not just a so-called objec-
tive material reality but is encompassed by innumerable
minds or psyches, including those of animals. The idea that
there is an objective reality, totally free from any kind of
psychic influence, is an extraordinary illusion from this
point of view. And yet that's the view that science has been
based on.

One corollary of what you're saying is this: if our souls
are all over the place, if we're projecting them out there all
the time, what an incredible invitation to be responsible to
take care of our souls, so that we are projecting the best of
ourselves—grace and blessing and not envy, resentment,
hostility. In other words, our morality is not a private matter. It's totally public. We've been saying that the other
creatures are grace-filled but that we can get less and less
grace-filled through envy and other issues. This seems to me
a new way of approaching morality. It's not about do's and
don'ts but about filling up our spaces in order to respond to

one another with blessing to blessing and with grace to grace.

I'd like to share one statement from the Jewish tradition, which is Jesus' tradition, about soul. Jews don't really have a word for soul as distinct from body because Jewish people, like native peoples everywhere, tend to be nondualistic in their origins. Here is one statement from a Jewish perspective about soul: "Soul is seen as an expression of a person's total state of being alive. The soul is a totality filled with power. This power lets the soul grow and prosper so it can maintain itself and do its work in the world. This vital power without which no living being can exist the Israelites called their *berakah* blessing."

Blessing is the Jewish word for soul. It is a lot like grace. So the question is, Are we blessing the blessed? Are we returning blessing to a blessed Universe? It's a new way to look at morality.

This recovery of the sense of the extended soul also fits with the rediscovery of the unconscious aspects of the mind. Descartes's view that the only kind of mind is the conscious mind was absurd, and ever since Descartes, people have had to reinvent the unconscious. Before Descartes, it was taken for granted that the conscious mind was part of a much larger psychic system.

Freud was only one of those who rediscovered the unconscious. The British writer L. L. Whyte in his book The Unconscious before Freud *gives a history of all the people, dozens of them, who have reinvented the unconscious since Descartes created this problem by saying that the mind is just the conscious mind. Again and again people had to say, no it isn't. Jung went further than Freud in saying that it's not just a personal unconscious but a collective unconscious.*

All humans are part of a collective memory on which we all draw. We're unconsciously connected with everybody else.

Moreover, we are also connected consciously. The sociologist Émile Durkheim at the beginning of this century made a central part of his study of society the idea of this conscience collective. In French, Durkheim's language, the word conscience means both conscience and consciousness.

These twentieth-century concepts of the collective unconscious and of collective consciousness are another way of telling us that our own psyche is extended, it's not just confined within our brains. Even our thoughts affect other people, and we in turn are affected by others' thoughts. My own hypothesis of collective memory through morphic fields expresses this same idea: what we learn and what we think can affect other people by morphic resonance.

Our souls are bound up with those of others and bound up with the world around us. The idea of the mind being inside our heads, a small, portable entity isolated in the privacy of our skulls, is extraordinary. No culture in the past has had this idea, and it's amazing that the most educated and sophisticated culture that has ever existed (as we'd like to think of ourselves) could have such an extraordinary view.

Habits, expressed through morphic fields, are an essential part of bodily and psychic life. Another essential part is creativity. Against the background of habit, creativity, freedom, and what you would call spirit, this creative living principle, are possible. If there were only habits, we'd become very repetitive and unconscious, because habit is largely unconscious. But we need habits because we can usually think about only one thing at a time. For example, the fact that we don't have to scan a vocabulary consciously for every English word while we're talking is very useful.

But we have to be philosophical about which habits we accept. The philosophical habit we've had for two hundred years may be quite weird, strange, and pathological.

Paradigms are indeed habits, they're habits of thinking. They become so deeply embedded that they become habitual for individuals and for a culture. Like the fact that ten- or eleven-year-old children in our culture accept that thoughts are inside their heads, because they're taught this in school and by their parents and it's a common assumption. This was an eccentric philosophical theory in the early seventeenth century. It's now a paradigm of our whole culture. These things get deeply embedded and become habitual, and we take them on at younger and younger ages so that it's harder and harder to examine them.

Just as we need to become aware of our scientific and philosophical paradigms before we can change them, so we have habits of emotional and social response that need to be brought to consciousness before we can let go of them. This is surely related to the practice of confession and to the Christian emphasis on forgiveness.

The topic of forgiveness is very important, one of the forms of letting go. I think we have an ideology in the West that forgiveness is altruistic. It isn't; it's our only way to get free. It's like remembering that grief begins with anger and rage. But the second level of grief is sorrow and the third level is transcendence—bottoming out. Many people are stuck in their anger, and to be stuck in our anger while in grief is understandable. But it means we're still in a servile relationship with the object of our anger. It's not doing us much good. We need to go past the anger and deeper into sorrow, and then even further into letting go. It's to our advantage—it's moving on. I think the metaphor of movement is important. We can freeze, but that is idolatry. Idolatry is freezing the object of your attention. We can make idols of anything from our objects of lust to objects of religion to objects of grief.

Aquinas has this great line. He says, "The first effect of love is melting." It's so sensual! The first thing love does is to melt. That's the opposite of freezing. We can be frozen with anger, resentment, hurt, and we've got to do some melting. Melting is always movement. I think that's a way to look at forgiveness too, as melting.

Again, it's not about altruism. I don't believe in altruism. I believe in compassion and love and the fact that when you love others as you love yourself, that's compassion. The love you put out there goes through your body. Remember, we're saying the soul is not in the body, the body's in the soul. We're not saying the soul is out there someplace unconnected to the body; rather, it's buzzing in the body, through the body, and beyond. At certain moments, we go beyond the body, but the point is that the body itself is being warmed or rendered cold by the soul.

Lack of forgiveness is a frozen state. It's interesting that you bring in the archetype of the tradition of confession in the Catholic church. The theory is a good one—let's forgive one another. But the practice limps. That brings us to the whole subject of ritual and how anemic ritual has become in the West. We have this rich theological tradition of sacrament but it's not producing for most people today. I think it's important how you say that you have found in your experience that we yearn for forgiveness. Who can teach us forgiveness?

I met a Croatian man a while back. He told me that in Bosnia, Serbs, Muslims, and Croats get along fine but their political leaders get into resentment. Resentment is lack of forgiveness. If you ask Serbs why they're torturing the Muslims today, they'll say, "They tortured us sixty years ago." They're right about that, but is this the way beings get on in the world? Is this why we're here? Who's teaching these people forgiveness?

How do you think ritual can help in the melting of resentment?

The first level of grief is rage. We need rituals that will take us into this rage so we aren't exploding all over the place and can get through it quickly. The native peoples have some interesting rituals for anger and for grief. One is drumming, drumming every day, beating the drum as hard as you can and wailing from the gut—that's where anger is, the third chakra. And so you just pour out your sound. What are you doing when you're pounding a drum? You're riding a horse, it's a horse's hide. You're riding a horse into the land of grief. I did this for a couple of months several years ago when I was going through some grief. It was a powerful meditation. I rode the horse into the land of grief.

Another simple Native American tradition is to take a rock and to pour your anger into the stone and wrap it in a bandana or precious cloth and bury it in the Earth. The cosmos can absorb our anger and our grief. That's why without a cosmology we don't deal with it. We need valium or a psychiatrist. But with cosmology we can return our anger to the stars, to the ocean, to the Earth. It's a very practical way of dealing with the Earth. The Earth is not just something pretty which exists outside of us. It is a healer. It's a priest and a priestess in the best sense of these words.

I like the idea of these rituals. Anger and resentment, like other patterns of emotional response, are not merely personal; they have a generic, habitual quality. For example, if you get into a resentful state of mind, simply by being in that state of mind you tune in by morphic resonance with countless people who have been resentful in the past, including yourself. So you are actually influenced by your own past resentments and the resentments many other people have felt. You tune in to a generalized sense of resentment. These things are transpersonal in the sense that they possess us. Good habits can also possess us. We're not usually very original. Most of the feelings, habits, states of mind we get into, many other people have had in the past. When we get into them we're linking ourselves to all the people who have been in similar states before us.

If we're in a state of grace, then we're part of what, in a Christian context, we call the Communion of Saints. We're linked with all those who've been in a similar state. But if we're in a state of resentment, we link with those who've been in a state of resentment. I think it's interesting that some psychologists, particularly James Hillman in the Jungian tradition, treat what Jung called archetypes not just as existing in the collective unconscious, but as beings that possess us. I would say that we tune in to them.

Our souls are influenced by the souls of others and by the collective psyches of our families, our nations, our religious traditions. They are largely habitual in nature. But they are also open to creativity.

This brings us back to the question of the relationship between soul and spirit. As I've discussed before, spirit is a more generic and a larger being than soul. In the Native

American traditions, spirit is another name for God—the Great Spirit is God. The Holy Spirit in Christian tradition is the Spirit that created and continues to create all things. Thomas Aquinas says that the very Spirit, the Great Spirit, the Holy Spirit, that hovered over the waters at the beginning of creation hovers over the mind and the will of the artist as he or she creates. In other words, that co-creation is still ongoing, the Great Spirit is still with us.

That Spirit is vast, and we don't control it. We call on it but it blows where it wills. It is God. It's another name for God.

Our souls are our animated beingness. This Great Spirit, God, not only works through soul and body but is present in all beings, even in beings that we may not readily think of as having a soul. We call these *inanimate* beings—such as stones, water, or even the air—but I think there's ample room for discussion here. Spirit is omnipresent but we cannot control it.

Spirit is breath. That's another way of seeing it. Hildegard says that prayer is nothing but inhaling and exhaling the one breath of the Universe, which is ruler or spirit. There's only one spirit but there are many souls.

Eckhart says that our soul is a sleigh that rides through every portion of our body. The body itself is a field. Then the soul goes wherever our mind and our heart go: all the way out to the one trillion galaxies, that's our soul at work.

Spirit, though, is greater than that. It's in the world and in our being but it's also beyond it. It's *the* Spirit, it's not *my* spirit. It's more than just our spirit, it's greater than that. There's a transcendence to Spirit. Because Spirit is, if you will, the divine soul. It's not ipso facto the human soul. We can banish Spirit or get close to banishing Spirit by our resentments and our choices in life. We can tell the Spirit to get out and then we become a home for small spirits, shadow spirits.

But then what about individuality? The soul is both individual and goes beyond the level of the individual; it's both/ and. We are responsible for our presence in the world. We have choices to make, that's what morality is about. We either make them or fail to make them. Even failing to make choices is a moral decision. Choices create our morality and our distinctness. You are distinct from your parents; you are also connected to your parents. It's both/and. You are distinct from your children and they'll show you how they're distinct from you as they grow up. But you're also connected. It's both/and.

Part of the accomplishment of the last three hundred years in the West has been to develop the sense of the individual. Unfortunately, we've done it at the expense of the community, at every level: the human community and the Earth community. That's what we're trying to crack. But I don't think we throw out everything that's been accomplished in terms of the individual. In other words, there is no need to return to the past when individuality was not honored. It is an accomplishment to be able to recognize the uniqueness that is special to each being, including each human being.

However, we've done it at a terrible expense. We have to get the dialectic back. But we should not go too far in the direction of ignoring the reality of individual responsibility and individual story. Even in our own families everyone has a unique story and that has to be honored. In one family we can have heterosexuals and homosexuals. They will live very different lives from one another, and that has to be honored. Girls come into this world with a morphic resonance different from that of boys because of the way we've been treated for centuries. Let's not muddle over the differences either. It's a both/and thing.

I would say that the individual soul is localized in the sense that the soul is centered on the body. It informs the body, it literally gives it its form. And the body is the center of action of the soul. The spirit is everywhere and also nowhere in particular. It's in all things. It's not localized in the same sense.

My soul, to do with my memories, my stories, my myths, my upbringing, and so on, is localized in my body and around it. The soul spreads around the body, but it does have a focus or a center. Souls are individualized in the sense that bodies are individualized. The point about bodies is that no two bodies can occupy the same space at the same time. That's how we define material bodies.

But that's not how fields operate. Different fields can occupy the same space at the same time. They can interpenetrate. The room you are in is full of electromagnetic radiations. Every radio and TV broadcast in the world is present, some weaker, some stronger, and with the right receiving apparatus you can pick them up. They're all interpenetrating. It doesn't mean you can't tune into some because others are present. At the same time, the gravitational field is permeating the room you are in. If it weren't, you would be floating in the air. Yet it's not interfering in any noticeable way with the electromagnetic field. Fields can interpenetrate.

Likewise, souls can interpenetrate, in the sense that a room full of people will also be full of their extended minds or souls. All people share the environment in this way. But the part of the soul that animates the body is localized and that's the center of each soul. The body is localized and our bodies can't occupy the same space, which is why there's a limit to how many people you can fit in a room.

Aquinas says that our soul is absolutely unique to our unique body. The important thing is that everybody is absolutely unique. However we wander into one another's souls, the home base, the body, is absolutely unique. My soul will never fit your body—which is really an interesting idea.

We should rise up and praise when we talk about what friendship is and love is and what lovers are about—the interpenetration of one another's souls by way of the body. That's so marvelous! I think angels are envious of humans because we have bodies and they don't, and lovemaking makes the angels flap their wings in envy. That's what the Song of Songs is about. Human sexuality is a mystical moment in the history of the Universe. All the angels and all the other beings come out to wonder at this. There's a Jewish tradition that the Sabbath day, Shabat, is celebrated in the act of lovemaking. You make love on the Sabbath. That's a necessary part of a day of thanks and rest.

CHAPTER FOUR

PRAYER

RUPERT

I'm going to start by talking about my own experience of prayer rather than about the theology of prayer, which I've never studied.

I prayed as a child, particularly at bedtime, having devout parents. This form of prayer was quite important to me; it was when I could ask God to help me in the things I was worried about, ask for protection and guidance. It became an important part of my life as a child. I had the idea that God could see everything, particularly that he could see through me or into me. I believed that God could hear when I prayed to him. I never heard God speak except when in chapel they said they were reading the word of God from the Bible. So I thought God's speaking was only in a book. But in my image of him, he could see and he could hear.

I wasn't particularly concerned whether God was he or

she or whether he had a beard or not, but the idea that God could hear was quite an early experience for me through the practice of daily prayer.

By the time I was about thirteen or fourteen I came under the influence of the scientific and materialist approach and realized that from a scientific point of view none of this made sense. I gradually became an atheist, believing that science had taken us beyond the superstitions and mumbo jumbo of religion, leading us into the light of reason, with scientists blazing the trail for a brighter and better future for humanity. In this materialistic world, prayer seemed to have little place or relevance.

The issue was brought to a head in my Anglican boarding school, when it was my class's turn to be confirmed. I refused to get confirmed; I was the only boy who wouldn't. I felt it was a false thing to do. But I went on praying for a while, and then I thought I'd try dropping it experimentally. When I had difficult days ahead I still prayed, and I stopped praying before days that I thought were going to be easy. Then I gradually and rather tentatively dropped the praying before difficult days to see if anything really dreadful happened, and it didn't seem to. So I gradually gave it up altogether and became more and more absorbed in a rationalistic and materialistic way of thinking.

I suppose I was still sustained by prayer because my mother certainly prayed for me every night and probably my father and my grandparents did too. But I didn't take that into account in my experimental test. And I still went to church services. I was in the school choir, and in any case the services were compulsory. In fact, I never completely gave up going to services. Even in my atheistic phase, I used to go to Evensong in the college chapel at Cambridge or in cathedrals because the music is so beautiful and the language of the prayers so moving.

It wasn't until many years later, when I was about thirty, that I became interested again in inner practices, having traveled to India, explored the rainforests of Malaya, and tried psychedelic drugs, notably LSD. The practice that made most sense to me was transcendental meditation. I came across people in Cambridge who taught it, and the way they explained it was very attractive to me. You didn't have to buy into a belief system, you didn't have to believe there was anything going on except calming the mind through a proven practice used for generations. That seemed all right; it didn't seem to challenge the scientific worldview because I could regard the whole thing as entirely internal and physiological. Indeed, the instructors rather encouraged this approach by showing graphs of how meditation could change the patterns of electrical waves in the brain and reduce blood sugar levels or raise lactic acid levels, I forget which.

They said to just try it, and I tried it, and it worked. I felt very calm. I did it for twenty minutes morning and evening, as recommended, for several years and found it very helpful. But it still didn't involve the dimension of prayer because I could regard it as a purely internal exercise. There is a strange way in which contemplative prayer or meditation or mystical experience is far easier for scientifically minded people to assimilate than petitionary prayer. Contemplative prayer or meditation can easily be regarded from a materialistic point of view as a purely internal operation, affecting the physiology of the brain by some kind of feedback process. It doesn't require belief in anything beyond your own nervous system. Petitionary prayer does. You're usually asking for things to happen at a distance, and that involves a much more profound challenge to orthodox science.

I went to live in India, working in an agricultural institute, and explored other forms of meditation, such as raja yoga. I met Sufis in Hyderabad, where I was living, and one of them became my teacher and gave me a wazifah, a Sufi version of a mantra, one of the ninety-nine names of God. I did that for about a year and I found it helpful. This form of meditation was centered in the heart, like many Sufi practices.

Quite unexpectedly, while I was in India I found myself being drawn back to a Christian path. When I found out about the Jesus prayer, in the Orthodox tradition of the prayer of the heart, I started practicing that, treating it as a Christian mantra. I still could think of this as an internal process and it didn't really challenge my scientific view of the world too much. But then I began, around the age of thirty-four, to pray again on a daily basis. I began by saying the Lord's Prayer every morning. And as I got into it, different phrases opened up for me as being particularly relevant and meaningful. I still do this daily, and when I'm praying with the Lord's Prayer sometimes one part of it can take a long time. For example, I may spend half an hour on "deliver us from evil," bringing to mind people I know who need deliverance from evil, the forms of evil that threaten me and my family, and those that threaten human society or the planet. Another time it may be forgiveness. So different parts of it can be the focus for petitionary and intercessory prayers, asking for things and asking for help for other people and for myself.

Meditation helped to take my mind away from mundane concerns; but when I came back, all the problems of the world were still there. I experienced a big gulf between the meditative state and the rest of life. In the Hindu tradition, this is one of the attractions of meditation. The purpose of meditation is not to transform the world, which is irredeemable, but to liberate the individual from the endless cycles of karma. The goal seems to be a kind of vertical takeoff, where you leave behind the world and all its concerns. By contrast, I found that prayer connected me more closely with what was going on in my life and in the world around me. I found prayer increasingly helpful, and gradually I began to pray more and meditate less.

Then I began to wonder about prayer and how it worked. Mechanistically speaking, the chemical and electrical changes that go on inside your brain are virtually undetectable a few feet away and are hardly likely to be able to affect the course of Nature or the course of other people's lives or to act on things at a distance. Prayer from this point of view is scientifically incredible.

I found very few people prepared to discuss this problem, even though surveys show that most people pray, even people who don't go to church. Prayer is quite common even in very secular societies like modern Britain. But for most people, prayer is a very private matter, and they would prefer not to have to think about it too much because it is so at variance with the scientific worldview. If you pray for somebody in Australia to get better and she does, how could that possibly work? Of course, if you ring up a person and tell him that you're praying for him, it could be psychological suggestion. Prayer might help you for psychological reasons. But for prayer to act at a distance without the element of suggestion defies all known scientific models of reality. It ought not to be effective, and yet most people do it.

When I got back to England, I suggested to my friends in the Epiphany Philosophers that we discuss the subject of prayer. They agreed, and we had a series of meetings spread over two or three years. I will briefly summarize some of the questions we discussed.

The first one was how prayer differs from positive thinking. Especially in America, there are endless books on how to achieve success in love and business through positive *thinking. These kinds of books have been published since the nineteenth century, and each generation has scores of best-sellers based on this concept. One is always meeting people who swear to the effectiveness of positive thinking, particularly salespeople, who often take courses in it. But most stories I hear concern getting parking places.*

The technique is to visualize and create a strong impression of what you want; then the idea is that it will happen. This differs from prayer in that you believe you are accomplishing what you want through the power of your own mind. It is not done in accordance with any greater purpose, it is just a method for getting what you want: getting the good deal that you want in business; getting the plot of land that you want to build your house on; passing the exam you want to pass; seducing the person you want to seduce. It is largely done for selfish motives. Positive thinking can be used in a wider context, but the way it is usually put across is for getting what you want through the power of the mind.

I think this can work. Moreover, I think the morphic field idea helps us understand how it could work. One can create a field of intention within the mind, and because the mind is not localized within the brain, this field can spread out and actually affect the course of Nature. Combining the idea of the nonlocal mind or extended mind (chapter 2) with the idea of fields of intention, one can begin to see how positive thinking might work.

But prayer is more than that. Prayer involves putting what you want in a larger context. Prayers begin with invocations addressed to higher powers: "Our Father, who art in heaven"; "Hail Mary, full of grace"; "Come, Holy Spirit, our souls inspire." Prayers in all traditions start with invocations.

In petitionary and intercessory prayer, asking for things and praying for people, there is a conscious alignment of what you are asking for with the higher purpose *"Thy will be done."* This takes prayer beyond mere positive thinking and involves more than one's own mind.

Second, in our Epiphany group we discovered that we were constantly at cross purposes in our discussions until we hit on the simple idea of asking the various members of the group to say how they themselves prayed. We had known each other and prayed together (following the Anglican prayer book) for at least ten years, but none of us had ever asked this question. It was astonishing how different the replies were. I was amazed that some members of the group would never pray for anything specific. They thought that it was wrong to interfere with God's will; you should just ask in the most general sense, *Thy will be done*, holding the person or the thing being prayed for in the light of God, as it were, and keeping the prayers generic.

Others said, no, what we're enjoined to do in the Bible is to pray very specifically. We ask for exactly what we want and it's precisely that clarity of intention, that sharpness of focus, that makes it effective. These differences present in our group are very widespread. In general, liberals favor generic prayer, while evangelicals are often very specific and can point to remarkable examples of how prayer can work.

I personally witnessed a striking example when I was living in Bede Griffiths's ashram in southern India. A young man was staying there who was training to be a pastor in the Church of Southern India and was from an evangelical tradition. He was very cheerful, full of joy, and trusted totally in prayer. He lived by prayer and prayed specifically for what he wanted.

One rainy day I was using a telescopic umbrella. He'd never seen one before and really wanted to have one. On the grounds that it would help him in his pastoral duties, he *prayed for one. This seemed to me an unlikely prayer to be granted because such umbrellas weren't sold in India at the time. The next day I saw him walking into the ashram with a smile that filled most of his face, clutching a telescopic umbrella. I asked where he had got it. He told me his prayer had been answered. He was walking along the road into the nearby town when an express bus rattled past. Just in front of him this umbrella dropped out of the bus onto the road, and the bus disappeared in a cloud of dust.*

To me this was a lesson in the power of specific prayer. No doubt the objection could be raised that this was more like an example of positive thinking, and it would indeed be difficult to make a very sharp distinction between the two. But those who pray very specifically are sometimes quite sophisticated. They are aware of the possibility of trying to use prayer for gratifying selfish desires but argue that by aligning ourselves with the will of God and praying specifically, we can become co-creators in the course of events. God himself is involved in the general direction of things, but forming the specific intentions is the very work that we're called to do as we pray.

Third, we faced the question of what to pray for. If we pray for the coming of the kingdom of heaven, what image do we have of it? We know what we're against: injustice, exploitation, pollution, the rape of the forests, and so on. But it's much harder to say what we're for, what we actually envisage, if we pray specifically for the coming of the kingdom of heaven and move beyond very general concepts like justice and peace. This is another issue where people differ considerably, and again the difference is between praying generally and praying specifically.

Most people are rather reluctant to talk about prayer. This is not so among evangelicals, and one of their great strengths and one thing I admire about them is their ability to pray spontaneously out loud. They regard this practice as perfectly normal. The charismatic movement has also given many people in nonevangelical churches a similar confidence.

Those are some of my reflections on prayer. I've concentrated mainly on petitionary and intercessory prayer, which are often regarded as the lowest forms. But I think that they are very important and are certainly the most embarrassing in the context of the present scientific worldview.

MATTHEW

Thank you, Rupert, for going beyond the usual coyness of the Anglican, to say nothing of the Anglican scientist, and sharing your personal journey. I think that's very valuable and courageous of you, even vulnerable of you. I was reminded by your story of giving up prayer at thirteen of how Huck Finn at eleven prayed for a fishing rod and hooks, and he got only hooks. So he gave up prayer for the rest of his life. It seems to be about that age when we reject one kind of prayer for another or for none.

I was drawn into the Dominicans for several reasons, but one was that I visited their priory when I was a senior in high school and the chanting of the office in Latin was so beautiful. It was the aesthetics that drew me in. I have much to thank the Dominicans for, and that was part of the training. The whole relationship between prayer and beauty I think is often underestimated.

The first book I wrote was on prayer, because I was very keen on spirituality and I felt that the number one issue was What is prayer? I felt through my whole training as a Dominican that we had prayer by osmosis but not by reflection. I'd been a Dominican about fourteen years when I

wrote that book. We were not really taught to be mystics, or
what it meant to practice mysticism. And what is prayer
anyway? My book with the unlikely title *On Becoming a
Musical Mystical Bear* was my book on what prayer is. I
still stand by the definition I come up with in that book
because I haven't seen a better one. My definition is that
prayer is essentially a radical response to life. It is our root
response to life. There are really two responses. One is our
yes and the other is our no.

The dialectic between yes and no makes up our radical
response to life, and therefore prayer is both our mysticism
(yes) and our prophecy (no). The struggle for justice is just
as much prayer as is getting high on the beauty of life. They
go together because the more we fall in love with life, the
more we are open to the vulnerability of the forces within
ourselves and within our society that endanger life: forces of
injustice, racism, and so forth. I found writing that book
very satisfying because I was struggling then, in the late
sixties and early seventies, with the issues of mysticism on
the one hand and the Vietnam War, civil rights, and
women's rights on the other.

We are at a moment in the human species when prayer
has to burst forth like a volcano. Think how much one volcano can do for the whole planet. We live on a crust only
twenty miles thick and below it is one thousand miles of
molten lava. I think our souls are the same way. I think
inside all of us there's this molten lava and it seldom bubbles
up. We go through life bored and die; everyone's bored with
us and life is boring. Prayer is the opportunity to tap into
that molten lava and let out those nutrients that the planet
needs, that the young need, that we need to be alive and
purified and aware. This moment of deep ecumenism is the
time for this to happen. By deep ecumenism I mean the
coming together of world religions and spiritual traditions

around prayer, around mystical practice, around mystical experiences entered into together.

I have two images of that. One is Meister Eckhart saying, "God is a great underground river that no one can dam up and no one can stop." And so we have these different wells into the river: a Buddhist well, a Taoist well, a native well, Christian, Sufi, Jewish wells. We all go down somewhere. We each have to practice some path in the nineties, and essentially these are paths of prayer. Going down the well we come to the one underground river, the living waters of wisdom. Prayer is a release of wisdom. That's the volcano that hasn't happened yet. We've got three hundred years of scientific knowledge; we invent computers to store all that knowledge, and how much wisdom have we developed in the last three hundred years? How much have we lost? What does it mean to wipe out aboriginal peoples willy nilly, if not to destroy wisdom?

Another image for this idea of deep ecumenism comes from Bede Griffiths. He talked about our five fingers. Let's say each finger is one of the world religions: Hinduism, Buddhism, Judaism, Islam, Christianity. When we look at the top they're quite separate. But if we look at the roots, they form one hand. To me that metaphor is very close to Eckhart's metaphor about the underground river. Both work.

The point is to get down the well and to cut the nonsense about the differences in our religions which are really so minor compared to what has to happen. Is there a volcano down there or isn't there? Where is the fire? Are there some nutrients to be dispersed or aren't there?

Nicholas of Cusa was a mystic and a scientist in the fifteenth century and a cardinal in the Roman Catholic Church (which isn't necessarily a very good reference). David Bohm, the physicist, said he owed more to Cusa than to Einstein, which I think is pretty heretical for a scientist to say. Cusa was the pope's legate to Greece and in the process of his work he learned something about deep ecumenism. He wrote this incredible statement in the fifteenth century: "Humanity will find that it is not a diversity of creeds but the very same creed which is everywhere proposed for there cannot but be one wisdom. Humans must therefore all agree that there is but one most simple wisdom whose power is infinite and everyone in explaining the intensity of this beauty must discover that it is a supreme and terrible beauty."

It is beauty and wisdom, awe and wonder that is going to bring about deep ecumenism, and I think Cusa's prophecy is emerging. In the very collapse of the forms of institutional religion we have evidence that something deeper needs to be born out of religion.

The four paths of creation spirituality name the four dimensions of prayer in a richer and deeper way than we've been told in our classical theologies of prayer. Very briefly, to make my point here: the first path of creation spirituality is the *via positiva* which of course is a classical term, not used very much. But it's about our experience of awe. Rabbi Heschel says, "Praise precedes faith." That's such an important sentence. In other words, without praise there is no faith. Praise precedes faith yet we cannot presume praise anymore in an anthropocentric urban-centered civilization like ours. There's no praise in a machine as such. We've lost the *via positiva*. I will always maintain that for westerners the *via positiva* is the most difficult of all the four paths. It's harder than suffering for westerners to get what praise is about and awe and wonder and the joy that goes with it. This is because we have denigrated the sources of awe: Nature and Creation.

This is prayer: rendering ourselves vulnerable to awe,

wonder, gratitude. It's the first stage of prayer. As Meister Eckhart says, "If the only prayer you say in your whole life is thank you, that would suffice." Gratitude, gratefulness comes from that.

The next path is the *via negativa.* The negative way is the way of darkness, suffering, silence, letting go, and even nothingness. Emptying. All these are prayer: experiencing silence; being emptied of images, verbal, oral, and imaginative; letting go; and suffering. It's not just about asking to be relieved from our suffering, it's about entering into the process, to learn. Suffering is one of the great teachers, one of the gurus of our lives, and the West is very bad on suffering. We want Valium or surgery or anything to block it or "make it go away." But the letting go process is prayer.

The third path is the *via creativa,* entering into the creative process not to produce a product but to honor our images by paying attention to them and giving birth to them. Thereby we honor our deepest experiences: our passions, our joy, and our sorrow, the cosmic Christ in us who is not only the light in us but also the wounds in us, in all beings. Creativity is itself a prayer process. One of my favorite passages about creativity is from Leonard Bernstein. He talks about how you can sit at a piano for hours trying to compose a piece and you're playing with the notes and the chords and nothing happens. Then all of a sudden, it happens, and the way he describes it is so beautiful. He says you forget all time and all space and you don't know where you are but you know you're part of something bigger than yourself. All you can say, he says, is *"Deo gratias."* Which is so fascinating coming from a Jew, because he breaks into the Latin Mass in the moment of being struck by the lightning of the creative process.

The fourth path is the *via transformativa,* the transformative way. This is a path of compassion which is about the

realization of our interdependence and the action that results from it. The first response to interdependence is,
"Wow we're all here, let's celebrate." The second response is
healing, which is another word for justice-making, because
the biggest ruptures are injustice.

All four paths are prayer. Our struggle for justice, our
yearning to celebrate, our creating forms that allow us to
celebrate entering into darkness and grief—all this is
prayer. It is a radical response to life because it comes from
the depths of our living.

Now I want to say a few words about meditation. There
is a group in England called the Study Society, and they
invited me to speak a while ago. For years they have been
practicing whirling dervish prayer. A very interesting
group—artists, philosophers, scientists, poets, and all. I
drew up a little reflection on meditation and what it does for
us. What do we learn from meditation? To be still; to be
empty; to be with being; to be with nonbeing; to be with
nothingness. Therefore we learn relationship from all of this
being-with.

The Lakota people when they pray always pray "Aho
Mitakye Oyasin"—all our relations. That's the essence of
prayer, it's purifying our relationships. Remember we said
earlier that the soul goes out, it's all-encompassing. There
are a lot of relationships out there to pay attention to. And
Eckhart says, "Relation is the essence of everything that
exists." That is today's physics! He said it in the fourteenth
century. No wonder he was condemned! He destroyed in
that sentence the twenty-five hundred previous years of
Western metaphysics: looking for the substance of things.
They're still looking for the thing inside the thing inside the
thing. But Eckhart said, "Relation is the essence of everything."

What if he's right? Then that would mean that the practice that gets into relation—purifying it, honoring it, praising it, and energizing it—that practice is the heart of things. I propose that practice is prayer, meditation. It teaches us to be with process. Is all relation process? Is all process relation, relation in motion? It seems to me that all beings are both relationship and process. Meditation teaches us to be with darkness, to be in the present. Fully in the present, which means to let go of the past and the future, all schemes, all projections, all projects, and all patterns. Therefore, it means to be open to the future pattern, to the not yet, to the unborn. Eckhart tells us to become our unborn selves. That's what happens when we return to the origins, we become our unborn selves. That is such an incredible invitation! Why? There's freedom. We can start all over. We have the capacity through meditation to return to our unborn selves. This is about learning to play, learning to laugh, to be youthful. Eckhart says, "God is *novissimus*"— the newest thing in the Universe. And if you return to God, you return to the beginning. This is why the Bible begins with the words "In the beginning," not only in Genesis but in John's Gospel as well. Because God is always in the beginning. Where are we? We've gone wandering. As Eckhart says, "God is at home, it's we who've gone out for a walk." Meditation is this return home to our origins.

We learn then to be at home in the dark and in the light, in suffering and in joy, in riches and in poverty. We learn surprises and openness to the spirit. We learn stillness and the silence that is behind all things. We learn to be with awe and to let it fill us. We learn from the God of darkness and chaos to be with chaos. The most important thing I can say about chaos is to trust it; it's all over the place now. Even the physicists are on board. It's certainly in our psyche. There is this effort by the human to want to control. That patriarchal morphic field that we've developed so well is especially about control. We are ill at ease with chaos, and we want to pretend it isn't there.

I'm convinced the number one male problem is self-pity. But what is self-pity? It's looking for the mother, the nurturing mother outside oneself. It's a search for the mother who was cast out of the male. The male has to find the mother inside himself and among ourselves. Men have to start mothering men. Francis of Assisi said to his brothers, "Be mothers to one another." Once when I proposed this in a lecture, a woman came up afterward and said, "Well you're probably right about men but let me tell you that one problem for us women is bitterness." It's interesting that Aquinas says bitterness comes from holding anger in for a long time. That fits because women have been told in our culture to be sweet, nice, and feminine and not to show anger. By being held in it festers, and it becomes bitterness. This prevents women from finding that power inside to express their anger and the forms to do it constructively and not destructively.

Meditation helps us in the sexual arena. I'm convinced we can't deal with sexuality apart from mysticism. It is always interesting to read the mystics on sexuality. I'll never forget the moment, fifteen years ago, when a woman first came to our institute and read Thomas Merton's Zen poetry. There's this image from Merton's poetry saying, "I am a bell," and I almost fell out of my chair because I thought, "That's such a feminine image!" Then I started to pursue it and I found that eighty-five percent of the images for the soul in John of the Cross, Meister Eckhart, and Merton are feminine. Then ninety percent of Teresa of Ávila's images for her soul are masculine. What is so great about the mystics is that they totally confuse us about gender. We think we're confused now about genders—wait until you get into the mystics! It's chaos, it's exactly what we need to heal the sexual problem in our civilization. We're so stuck on what maleness ought to be and what femaleness ought to be that we need to go into the level where meditation can take us and let sexuality be sexuality and let it emerge as it yearns to express itself in whatever those expressions be.

We also encounter in meditation the God of compassion calling us to be other Christs, indwelling, lights of the world and inheritors of the divine "I am." All this happens in meditation.

But I also want to talk about two different kinds of meditation. One is the active where we're told to imagine an image, to put out an image, or we're given an image to follow. In active meditation in some healing practices you are instructed to meditate on your cancer: you fight it, and you go in there and become the white blood cell and so forth. That has a place. It is fine, but it's not the only kind of meditation.

The emptying meditation, the receptive meditation, is terribly needed today. This is found in Zen and in other practices too. The emptying. The simple instruction is silence — listen to our own image. That is the emptying — entering into our non-images, which is the moment of creativity. It's that thin line between nothingness and something new. Eckhart says, "I once had a dream that even though I was a man I was pregnant, I was pregnant with nothingness and out of nothingness God was born."

The promise from the mystics is that what we generate in emptying meditation is nothing less than the cosmic Christ. We generate the God-son and -daughter. As Eckhart says, "What good is it to me if Mary gave birth to the son of God fourteen hundred years ago and I don't give birth to the son or daughter of God in my time and in my culture and in my person."

The prayer process is a creative and generative process. It

never stops with the *via negativa* alone. Art-as-meditation, which follows on the emptying process, is a very useful, practical, and enjoyable form of prayer. It's delightful. I can tell you after eighteen years of teaching in our institute and using art-as-meditation that it's one of the most basic forms of education and prayer. The incredible conversions that happen in people's lives! I have learned that there are all kinds of artists in our world who have covered up their artistic talents. It's a volcano just waiting to erupt. Meditation lets it out much more swiftly than art classes and techniques do. As M. C. Richards, world-famous poet and potter, said at age seventy-two when she took up painting: "It's too late for technique." I think we could say that about our civilization: "It's too late for technique, let's get on with art." Technique may take care of itself as we go along.

The experts on prayer are the mystics, and I can't encourage enough the reading of them with our right brain and not our left brain. How do you do that? You read a mystic with your right brain in a bubble bath or in bed with your lover or at your favorite stream or among flowers! In other words, in a very sensuous place where the left brain cannot do its usual cut-'em-up work. You just put your left brain to bed and you find a sensuous place. As you read the mystics, the purpose is not to finish the book. It's to stop whenever you're struck by the wonder of it, the truth of a passage, the beauty of it. Just stop and let that image or that phrase wash over you for as long as you want. You referred to this, Rupert, when you told what you do with the Our Father, spending twenty minutes in one morning on one phrase.

The mystics are poets of the soul and they write in images. Every one of them was involved in language redemption. Julian of Norwich has been called the first woman of

English letters. Francis of Assisi was one of the first writers in Italian. John of the Cross was one of the first writers and one of the greatest poets in Spanish. Meister Eckhart was the first preacher in German. At his trial, one of the reasons he was condemned was that he was preaching to the peasants in their own language, the German dialect. He was accused of confusing them, telling them they're divine, disturbing the status quo. Don't underestimate that connection between redeeming language and what the mystics do for us.

I want to offer a few examples from Rumi, what he says so well to some of our Western mystics. You'll see the point I started with about deep ecumenism, the underground river. When we get into prayer, it's not that big a deal. Here's a great poem about praise called "What a Blessing":

Don't hide.
The sight of your face is a blessing.
Wherever you place your foot there rests a blessing.
Even your shadow passing over me like a swift bird,
   is a blessing.
The great Spring has come.
Your sweet air blowing through the city, the country,
   the gardens and the desert, is a blessing.
He has come with love to our door, his knock is a
   blessing.
We go from house to house asking of him, any
   answer is a blessing.
Caught in this body we look for a sight of the soul.
Remember what the prophet said, one sight is a
   blessing.
The leaf of every tree brings a message from the
   unseen world.
Look, every falling leaf is a blessing.

I remember hearing a woman interviewed on the radio who had been blind all her life and her sight came back at the age of forty. She was asked what was the first thing she did. She said, "I went and I watched a leaf fall." I said, "My God, I've had eyes for forty years and I've never taken the time to watch a leaf fall."

> All of nature swings in unison, singing without
>   tongues, listening without ears, what a blessing.
> O soul, the four elements are your face.
> Water wind fire and earth, each one is a blessing.
> Once the seed of faith takes root it cannot be blown
>   away even by the strongest wind, now that's a
>   blessing.

Rumi continues and then ends the poem this way:

> The heart can't wait to speak of this ecstasy.
> The soul is kissing the earth saying O God what a
>   blessing.
> Fill me with the wine of your silence, let it soak my
>   every pore.
> For the inner splendor it reveals is a blessing, is a
>   blessing.

We see that blessing consciousness is not owned by the Jewish tradition or by the West. This experience of blessing, the *via positiva,* is absolutely transcultural. Here's a wonderful line from Rumi.

> On the night of creation I was awake, busy at work
>   while everyone slept. I was there to see the first wink
>   and hear the first tale told. I was the first one caught in
>   the hair of the great imposter.

I don't know what that means, but it's fascinating.

> Whirling around the still point of ecstasy I spun like
> the wheel of heaven. How can I describe this to you,
> you were born later. I was a companion of the ancient
> lover, like a ball within a broken rim I endured his
> tyranny. Why shouldn't I be as lustrous as the king's
> cup. Sssh, no more words, hear only the voice within.
> Remember the first thing he said was, we are beyond
> words.

So there's the *via negativa*—"we are beyond words." But this imagery, being with God before the creation, is straight from the book of Sirach in the Bible. It's just like wisdom who was there before the creation of the world.

When you get into deep ecumenism, you're talking about our common experience. Nothing is more common than prayer; you were saying this at the practical level—everyone prays but no one talks about it except Evangelicals and their prayer is sometimes a little off center.

> We search for him the beloved, here and there while
> looking right at him.
> Sitting by his side we ask, O beloved, where is the
> beloved?

It's like John at the Last Supper asking Jesus, "Where is the father?" And Jesus says, "I've been with you all this time and you haven't known him?"

Rumi says, "Enough with such questions, let silence take you to the core of life. All your talk is worthless when compared to one whisper of the beloved." So there is that whole sense of being present to the beloved.

"Without looking, I can see everything within myself."

Eckhart says all the names we give to God come from our experience of ourselves. It's true, if we are out there in the Universe that's how we know God. There's this net that we throw into the Universe and then we gather in images for divinity. Rumi says, "Why should I bother my eyes anymore now that I see the whole world with God's eyes." Eckhart says, "The eye with which I see God is the same eye with which God sees me." That's what Rumi's said. We're catching eyes; it's eye to eye here, divinity and us.

Rumi says, "Do not look for God, look for the one looking for God. But why look at all, God is not lost, God is right here, closer than your own breath." That's just like Eckhart saying, "God is at home. We're the ones who have gone out for a walk."

We're totally undereducated in the West about what the mystics can do to help us pray. The key thing is that the most important mystic in your life is you. I don't celebrate Hildegard and Aquinas and Eckhart as mystics to do them some kind of homage. I celebrate them for the same reason that one learns to play tennis with someone who knows how to play tennis. I want a challenge to move beyond religion to be good pray-ers and to do that I go to the expert pray-ers, the heart specialists, the poets of the soul, the mystics. Why fool around with the third-class stuff when you've got all this first-class stuff available? Now we have it from East and West. The reason to love Hildegard and Eckhart and Rumi and Jesus as mystics is that they can bring the mystic alive in us, and that's what's so important. Without the mystic coming alive we're not going to have the imagination or the courage or the energy to be instruments for compassion and for the environmental revolution.

Wonderful! You have such a broad view of so many different kinds of prayer it's difficult to know where to begin.

But since I'm a scientist and since this question preoccupies me, I'd like to ask you how you think it works. How do you think these various kinds of prayer work? It's easy for us to say that they affect the person praying. It's easy to see how they can affect other people, if the other people know you're praying for them. But in the larger context, how do you think of prayers as acting?

That's an interesting question. I quoted Hildegard earlier saying that no warmth is lost in the Universe. My phrase is that no beauty is lost in the Universe. If I get it right, Einstein is saying that no energy is lost in the Universe. So I think the gathering up of our awe and wonder and suffering and creativity and compassion, all this gathering up, this volcano, this spewing out, goes out there just like the nutrients do in a volcano. Spiritual nutrients. They are not material so they can go very far. In this context we are beyond space and time where our prayers gather the beauty, the warmth, the compassion, the suffering, the awe, and the creativity of our ancestors. And they fertilize that which is to come.

We can imagine the soul as being out there already. There's this great being and of course, as you pointed out, we're all intersecting. The image I have is of bubbles. We're bubbles that intersect, transparent with one another. Our souls are these huge bubbles. We're already inside one another, so let's get on with it. Of course, that's a physical fact too; we're breathing in one another's water vapor but we're already inside one another's joy, we're already inside one another's suffering. That's why your point is well taken about how coy we are — we're pretending that we're so modest about our souls, we keep them covered all the time, but it's self-defeating. Eckhart defines compassion in this way; he says that what happens to another, whether be it joy or sorrow, happens to me. That's a law of interdependence but that's compassion. I am in your joy and I am in your sorrow. Joy and sorrow are not private property.

So does that explain how prayers happen?

Yes, it does in a way. What you're taking for granted is what I spent a long time struggling to arrive at, namely, a concept of the extended mind or the extended psyche, and this is where we found in our discussion of the soul that we'd come to similar conclusions (chapter 2). What I called the contracted mind, through my historical analysis of mechanistic science, you referred to as the shrunken soul.

You're taking for granted the idea of an extended psyche and a spiritual reality that goes far beyond our bodies. Therefore what we do in our psyche and our minds and through our prayer can have distant effects because our psyches and the spiritual reality in which we're embedded are nonlocal, they go beyond the limits of our body. This seems to be your presumption. I agree with it, but it's taken me a long time to arrive at that conclusion. I'm surprised at the ease with which you assume it.

What released me into it is where Eckhart says, "The soul is not in the body, the body is in the soul." He's right. That's my experience and I bet that's your experience too.

Yes. Also, the soul is not merely personal but collective. Positive thinking can be collective too—it's practiced by Japanese workers in car factories, for example. But how would you see the spiritual dimension as going beyond the merely psychic?

You often surprise me because you get me going in your images and then your question is different from what I thought was coming. I want to answer the question you didn't ask. I love your logic, it's chaotic to me. Let me give you my answer and you can guess my question.

My answer was going to be that there's this immense shadow side to mysticism. I immediately thought of Hitler. He used mystical practices: chanting, candles, the darkness of night, marching, camaraderie. All these under Hitler I call pseudo-mysticism. I feel that one reason Hitler was so successful was that the churches were doing so little to develop healthy mysticism. Mysticism is so endemic to our species that we yearn for it one way or another.

So that's my answer to the question that I thought you were going to ask me.

The question I did ask, just to remind you, is how you see the difference between the psychic and the spiritual realm, the difference between personal or collective positive thinking and what you regard as true prayer. Or maybe you don't make a sharp distinction between the two.

I don't make an ironclad distinction. I think it's a question of growing those souls, making them larger. You use the word *selfish* and I don't use that word a lot because I think it's been overworked in religious circles. As you know, I don't believe in altruism. I think all authentic love is self-love, but it's healthy self-love. Some New Agers do try to live in the light all the time. They're not living in the shadow, their own or someone else's. One way into the collective is through suffering, through that shadow, so I encourage them to go deeper. It's not to belittle their psychic needs but to show them that their psyches can grow much bigger. It's like what you said earlier, their psyches are smaller, so they've created a prayer for that little psyche. To somehow blast through that psyche I think the issues today—the ecology crisis, children's despair, children leaving their religious traditions—have the ability to wake people up.

It's a question of enlarging the psyche so it can become more cosmic in size. That includes not only dealing with the Christ of light but the Christ of wounds. Again Eckhart: "Our soul grows by subtraction, not by addition."

What does that mean?

Letting go is subtraction. It is simplifying our lives and learning to let go instead of always adding on. I think the Western habit is to add on. Suffering teaches us to let go, doesn't it? One route to subtraction is suffering. Another route is meditation, where we become simply present. Even sound, not taking sound for granted—that's a subtraction. Instead of adding all kinds of words, we need to get back to the essence of words, which is sound. There's a stripping down there and I think that's a soul growth process. A simple mantra chant can be a subtraction.

But I want to ask you, How do you see the definition of prayer, the practice of prayer, shifting from the Cartesian age to the new paradigm?

In the Cartesian paradigm prayer is subjective, and what happens in the real world is objective. The objective world is the realm of science, and prayer can have no objective effect. That is the old paradigm view. Those who believe that prayer works do not usually want to get involved in arguments with scientists, so they keep quiet about it, treating it as an intensely private matter in the realm of faith.

In the new paradigm the division between subjective and objective, inner and outer, is broken down. Our minds, our psyches, extend all around us. Then we are part of a collec-

tive psyche, with a collective unconscious. In this context, the extended effects of prayer and the extended effects of Spirit become much more possible, indeed likely. Prayer then seems less subjective. The division between the subjective and the objective is blurred. Prayer is no longer simply confined to the inside of your skull. The whole subject of prayer is much more credible. Because it is more credible it is no longer necessary for everyone to be so coy and secretive about it. The context for prayer will change as the new paradigm replaces the old.

And worship wouldn't encourage this coyness. We would encourage a certain extraversion in spiritual practice.

That's right. In fact, in the churches we have inherited prayers and forms of worship from a prescientific age that take for granted the efficacy of collective prayer.

Yet I would say if analyzed from a theological point of view, most of those prayers are theistic prayers and not pantheistic prayers. They're not leading the mystic out of the people that much. The whole idea that the written word, the Bible, is where God especially speaks is a prejudice we've had since the sixteenth century. I think we're losing that today. Now we're returning as a culture to images rather than words. In many ways this is a premodern movement. Ours is becoming much more of an image culture. This is why I have hope and even faith in the generation of the twenty-year-olds because I think they're moving beyond the word *culture* that actually antedated the scientific revolution. Beginning in the sixteenth century, religion built much of its worship and theology on human words. The invention of the printing press had much to do with its modern and textual approach to things. But now it's another era,

and images are closer to the mystical experience than words
are. This may be useful, it may be part of the hope of our
time that we're going to become more image-aware, but not
the imagery of the television networks. What I mean is listening to our own images more, and truly bringing out the
artist in everybody, bringing out the artist in the community. The community as artist.

CHAPTER FIVE

DARKNESS

RUPERT

Darkness is the polar opposite of light. Light and darkness are correlative terms, and there's a polarity between them.

There are two senses in which light involves darkness. The first is perhaps the less interesting and flows from what we know of the spectrum of electromagnetic radiation. The part of the spectrum we can actually see is a very narrow band, the visible spectrum. Then we go over into the infrared and radio waves at longer wavelengths, and the ultraviolet, x-rays and gamma rays at shorter wavelengths (Figure 5.1). There's a vast amount of radiative activity going on, every room is full of all sorts of radiation which we can't see. The great majority of what qualifies as light in the general sense of the word, electromagnetic radiation, is invisible to us. In that sense it's dark.

But when we speak about true darkness we're speaking about darkness which is the opposite of light. We can see that point immediately if we look at the very organ with which we see, the eye. The center of the pupil is black because the inside of the eye is black. You can see light because that which receives the light is dark. If the inside of the eye weren't black, the light that went in would scatter all over the place and we wouldn't be able to distinguish the light that was coming in from that which was reflected within the eyeball. Cameras are painted black inside for the same reason. It's the contrast between the blackness of the eye and the incoming radiation that enables us to see.

In fact, darkness is contained within light itself. This is shown by the phenomenon of diffraction. If you have two slits, for example, and the light goes through the two slits, then when the light rays interfere you get a series of light and dark bands (Figure 5.2). These light and dark bands tell us that light travels in waves.

Waves can be represented on graphs, as in Figure 5.3. The part of the wave above the horizontal line is light and the bit below it is, as it were, negative light or darkness. In waves of light there is exactly the same quantity of negative light as there is of positive light, just as much below the line as above it. In this sense darkness is an active principle, the polar opposite of light. Light is a vibration in the electromagnetic field, and this field itself is inherently polar because electricity is polar, with positive and negative poles, and magnetism is polar, with north and south poles. Electromagnetic radiation involves an oscillation between positive and negative, and north and south, which is expressed in the alternating pattern of light and darkness that shows up when light is diffracted.

Darkness is paradoxically contained within light in the same way that silence is paradoxically contained within

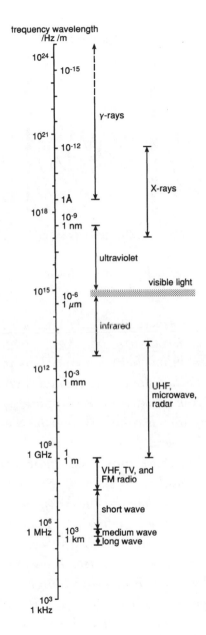

Figure 5.1. The electromagnetic spectrum.

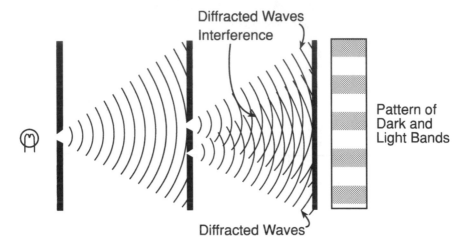

Figure 5.2. The diffraction of light. After passing through two adjacent slits, the light waves interfere with each other and produce a pattern of light and dark bands.

sound. Sound is also a vibratory wave phenomenon. When you clap out a rhythm, between the claps there's a silence. And it's precisely that silence that enables you to distinguish the sounds. If you speed up the clapping, speed up the vibration, it turns into a tone. As you further speed up the vibration (or frequency), it turns into higher and higher pitched tones. But those tones, like light waves, involve an alternation of sound and silence. So sound has silence within it, just as light has darkness within it.

There is also a sense in which the polarity between light and darkness has a historical origin, according to modern cosmology. At the time of the Big Bang, the Universe was very small and was at an incredibly high temperature. It is sometimes described as a fireball, but it wasn't fire in the normal sense of the word, in that fire radiates light. So peculiar was the state of this highly condensed Universe that there wasn't any differentiation between radiation and matter. Matter hadn't yet condensed out. It was so hot there

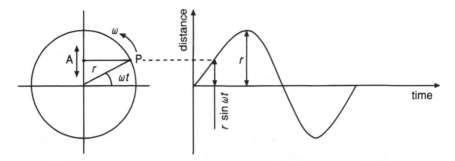

Figure 5.3. A sine wave and its mathematical relationship to circular motion. The point P moves around the circle at a constant speed, represented by the angular velocity ω. Point A—the foot of a perpendicular from P onto a diameter—moves up and down with simple harmonic motion. The up-and-down movement of A plotted against time is called a sine wave because the equation describing this motion involves the sine of the angle *t*.

was no such thing as matter as distinct from anything else. It was "a burning world of darkness," as the physicist Heinz Pagels put it. A similar state still exists in the interior of the sun. But as the Universe expanded, it cooled down, and after some 300,000 years it reached a point which physicists call the decoupling of matter and radiation. It was now cool enough for atoms to form, and all at once the Universe became transparent, bathed in a brilliant yellow light.

This primal differentiation of light and darkness recalls the description in chapter 1 of the Book of Genesis. The primal creative acts involve the establishment of divisions, first of all between light and darkness. Modern physics in its creation story also tells us that there was a primal undiffer- entiated unity which then underwent a series of progressive differentiations through this splitting apart of polarities. Even the fields of nature are supposed to originate from a primal unified field by a process called "spontaneous sym- metry breaking."

The primal division between light and darkness was a

division between light and atomic matter. Matter is that which can intercept light, absorbing it. And light is that which can move freely between matter. As soon as matter exists and can absorb light, there can be shadows. There can be the kind of darkness we experience in deep caves, or at night, when the matter of the Earth is absorbing the light of the sun. Ultimately, this division between matter and radiation underlies the distinction between night and day.

The daily rhythm of light and darkness, of day and night, is clearly one of the primary facts with which plants and animals have to deal, and it means that at night they do different things than in the day. Plants in the day carry out photosynthesis, and in the night engage in different activities, such as growth. Of course, in the animal kingdom there are daily cycles of activity which we know as sleep and waking. Except in the case of nocturnal animals, sleep is associated with the daily cycle of darkness. Our own experience of darkness through sleep has a biological history going back hundreds of millions of years.

The cycles of day and night, of waking and sleep, are also related to the polarity between our conscious and unconscious minds. Our minds are not conscious in the normal sense when we're asleep. What exactly is the state of the mind when it's asleep? The body goes on functioning, the heart goes on beating, we go on breathing. But our mind loses its normal consciousness, and consciousness is associated with light. We talk about the light of consciousness, about the inner light in our consciousness. In our dreams, arising from the sleeping mind, we experience the activity of the unconscious, which Freud and other psychotherapists have spent so much time thinking about.

Finally, when considering darkness from the scientific point of view, we need to call to mind dark matter. It's been found in the last ten years or so that the vast majority of the matter in the Universe is utterly unknown to us. It's invisible and undetectable by virtually any means except gravitation. As matter it has a gravitational effect. It's thought to exist because galaxies interact with each other in a way that shows they're much more strongly attracted to each other than they ought to be considering the total amount of matter in their stars and making a generous allowance for planets and even for black holes. This "missing mass" or "dark matter" turns out to constitute ninety to ninety-nine percent of the matter in the Universe. The cosmos seems to be grounded in dark matter, but no one knows what this substance is. This is currently a fertile field of speculation in theoretical physics.

Through the discovery of dark matter cosmology has, as it were, recognized the existence of the cosmic unconscious. The kinds of matter and energy that we know about are floating on the surface of a vast, unknown ocean of darkness—just as the conscious mind floats on the unconscious mind.

There's quite a close parallel here with the breakdown of Cartesian psychology, where the only kind of mind that was supposed to exist was the conscious mind. Before Descartes, as I've said, no one ever thought that the conscious mind was all there was; the rational mind was embedded within a much larger psyche, largely unconscious, linking us to the vegetative and animal realms, as well as the realms of emotions and dreams. But Descartes denied this larger psychic reality, confining the mind to the rational intellect. Consequently, the unconscious mind had to be reinvented again and again, most notably by Freud and Jung. Likewise, in cosmology it was thought by rationalists that we knew all there was to know about the matter in the Universe, or if we didn't we soon would. Now it seems that what we know about is really only a tiny fraction of what's there. Dark *matter underlies everything. It shapes the structure and will determine the fate of the cosmos in a way we do not, and perhaps cannot, understand.*

MATTHEW

This has me thinking about the relationship between cosmos and psyche because, as I hear you speak, I keep hearing resonances of the whole mystical tradition. As Eckhart puts it, "The ground of the soul is dark." I am reminded of the mystical tradition that the Godhead is dark, that it cannot be seen by its actions. The Creator, Redeemer God is active and the results are visible. But the Godhead from which all things begin is supremely dark. It's "a not-active superessential darkness that will never have a name and never be given a name." It's the great mystery.

In many ways, what keeps coming back in what you're saying here is mystery. The whole Universe or ninety percent of it is a dark mystery. This is like our psychic life, which comes through once in a while with memorable dreams, as you say, at night. I always ask, "Where do dreams come from?" I recently asked a good friend of mine who wrote a very powerful book on dreams: "Without using the word *unconscious,* tell us where dreams come from." I think dreams come from the Universe.

I'm just responding in terms of what I touch into through your meditation on this powerful light/dark theme. As you indicate, it's the first section of Genesis. It's also the prologue to John's Gospel: the Christ who is light comes into the world and the darkness does not overcome it. It's also the coming of the Christ at the dark time of the year, Christmas. All of these archetypes are inescapable in the Western tradition and indeed, from what I know, all spiritual traditions.

Another theme that is translatable into mystical language is that of illumination—the idea that we are illuminated, that we are enlightened, that the light in us increases and, paradoxically so, as we go more deeply into the dark, as we sink. This is the *via negativa* theme of the mystics, that illumination comes at the end of the bottoming out in the darkness experience. The psychic experience of the dark night of the soul and suffering as darkness, something that we do not have control over, something we can't really name but which is big within us, the great mystery again. Our culture would try to intervene with quick remedies, whether it's drugs or palliatives of some kind. A lot of our addictions are efforts to intervene with the darkness that's happening. But the mystical traditions would all say there's something deep to be learned by making the journey into the darkness.

One point that comes to my mind through your opening to this chapter is—and I love this analogy—silence is to sound as darkness is to light. It's so important, especially in the West where we tend to honor sound and light at the expense of silence and darkness. But my understanding is that the Big Bang was no bang at all, that it was utterly silent at that time, that there was no noise. Is that accurate, is that how you see it? That there was a great silence, not a big bang? I guess there was no air and therefore nothing to carry sound.

The Big Bang was a name given to this cosmological model by the astronomer Fred Hoyle, who was, strangely enough, also its deadliest opponent. Fred Hoyle was in favor of the continuous creation theory, and he's trying to stage a comeback by pointing out problems with the Big

Bang model. He used the term Big Bang as a pejorative term for this cosmological model. But it stuck. For the primary cosmogonic event to be compared to the explosion of a bomb is not very appropriate.

There must be some kind of Pentagon influence there.

There was recently a competition in one of the astronomy magazines for people to come up with better names—but none have caught on. For the primary event of our modern creation story to be named so derisively is not good. It's also not exactly accurate, as you say, because a bang implies acoustical waves, which isn't the right way to think about it.

Another point you made was about what you call the invisible light or dark light, rather than just darkness. What I like about that is the realization that there's much more to light in the world, just as there's much more to sound than we can pick up. We underestimate the amount of light in the world; as you say, in any room there are all these light waves bouncing around that we don't pick up. But also we underestimate the amount of darkness. It's interesting—we see again that our worlds are so small, that our psyches are so much smaller than the reality of the cosmos. What an invitation for people to always be on the alert to be learning, to pick up more lights, more sounds, a greater frequency but also more silence and more darkness. So maybe those four things are what learning's about.

I would like to return to the mystical celebration of the *via negativa* of the God of darkness and of our own spiritual darkness, which is not necessarily exclusively suffering. For example, it can be silence. Some mystics recommend the letting go of all images: images imply light; we see images only in the light, and images in some way reflect light so they all are luminous in some way and hopefully even numinous from time to time. One method, one path, into spiritual depth is to let go of all images, even our most cherished ones, including all visual images and audio images, and to be sinking into the silence.

Interestingly enough, Mechthild of Magdeburg speaks of "sinking and cooling," in the sense of the setting of the sun. The sun sinks and cools just as we do at the end of the day. We need to get this rhythm back again that you talked about, that we mammals do different things at night than we do in the day. This is where I think many New Age efforts at spirituality fail because they're so taken by the light and so eager to hold up light and warmth in a world that can be cold and dark that they don't honor enough the darkness, the sinking, the suffering, and the shadow.

It is helpful that you brought that out too, the relationship between light, darkness, and shadow, and that you honor them all: the idea of cosmic Christ as a light, but also the shadow and the divinity—how divinity comes in a shadow. The crucifixion was a time of shadow; the shadow of human violence and injustice showed itself on Calvary, executing a very good man. One of the symbolic ways of expressing this was that the sun went down at midday, so the whole cosmos was affected by this act of suffering. How archetypal this is not only for us humans maybe but for the entire cosmos itself.

But it is also redemptive. This sheds light on the meaning of the redemptive dimension to the Cross. One dimension of its being redemptive is that it reminds us of the journey every day into the utter darkness, the eclipse of the sun, which is how Good Friday is actually described in one Gospel. Then of course we have the luminosity of Easter

Sunday morning, of the sun rising and the Christ rising together. The sun energy returns again. Very cosmic, isn't it? Not unlike the birth story at Christmas in which the Christ came in the darkest time of the year.

Yes. And of course there is also the darkness of the tomb, the cave, and the descent into hell, the underworld, followed by ascent into light. These themes are fundamental in many shamanic traditions too—the journey downward and the journey upward, into darkness and into light.

The journey in and the journey out.

Yes. It's a paradox that when people actually go into darkness and stay there, in dark caves or sensory isolation tanks, after a certain amount of time, which may be only a matter of hours, they start having vivid hallucinations, vivid visual images—which are light.

When mystics talk about going into the darkness, I wonder if this is one of their meanings, because being in a literally dark place leads to visionary activity. This is one of the appeals of sensory isolation tanks. There are various spiritual practices in the Tibetan and other traditions that involve going into darkness and seeing visions.

When mystics speak of going into the darkness, do you think they mean literal darkness or metaphorical darkness?

Most of the time I think they mean psychological darkness, letting go of all images. That's how Eckhart talks about it time and time again. It's really about ceasing to project and ceasing to give birth to images. Again, I think part of it is the womb imagery too, the darkness of the womb, and out of that new images will be born eventually.

I don't think they mean so much turning the lights off in your room as turning the lights off in your mind.

The curious thing is that there are lights in the mind. For example, when we dream things seem to be lit up, there's a kind of internal light within the mind.

Yes, I think all images imply light. This is where Eckhart talks about the non-images and he talks about letting go so as to experience the "non-God." The idea that God is part of the projection, of the image of light that we have in our minds. There's a time for the darkness and the light, just like your wonderful image of the waves that go above and below equally. The above would be God and the below would be Godhead.

To get back to that kind of rhythm again would be to allow oneself to be as at home without images as with images. If we ask the question "What is our culture doing for us?"—it's constantly feeding us light and sound through television images and advertising images and music. We're so overdeveloped at the top level when walking around our culture that you have to ask, How can we develop the bottom? Certainly that's where going into Nature again can help. At least you rediscover the night as it is, the night as dark. Whereas without being outdoors in Nature we have lights on until the very moment that we go to bed and go to sleep. We're not giving darkness its due.

But when the mystics, like Eckhart, talk about going to the darkness and being free from images which imply light, how does this differ from the kind of consciousness that we have unconsciously, as it were, during deep sleep? There's a sense in which all of us descend into the darkness and let go of images every night. In various Asian traditions there's the

*idea of entering into a meditative state in which you remain
conscious in imageless sleep.*

*Do you think that what Eckhart is talking about has par-
allels in Hindu and Tibetan meditative practices where they
learn to go into this state of consciousness in deep sleep? I'm
not suggesting that he was necessarily going to sleep when
he was doing these meditative practices, but there must be
some connection between the state he's talking about, going
into the darkness, and the way in which some people can
retain awareness in deep, imageless sleep.*

Yes, I agree that there must be some kind of connection,
and one thing I hear when you speak that way is how rich
this area is and how unexplored in Western psychology. The
doors into the various rooms of the states of darkness and
imagelessness are much richer than we can imagine.

I would say that there's a big difference between the
darkness of sleep and dreams and the darkness of meditation. Maybe dreams come halfway, a kind of shadow between that and the radical letting go of all images that Eckhart was saying we are capable of in meditation. I think the
exact expression of Eckhart is found more in Zen than in
any other Eastern statements that I've seen, this letting go of
all images. When he says, "I pray to God to rid me of God,"
he knows our capacity for emptying.

Of course this brings me back to the image of the eye.
You say that it is black inside, but isn't that empty too? That
is nothing, it's no thing, and therefore it can be all things.

*The retina is black because of melanin, the pigment that's
in the skin of black people. It's very heavily loaded with this
black chemical.*

There's also a hole there, isn't there?

Well, the light goes in through the hole, the pupil. What you're seeing when you see the black in someone's eye is the black pigment in the retina.

That's terribly important to the mystics, to get not only into the blackness but into the hole, where we experience the bottoming out of all presuppositions in order to become utterly receptive. That is of course how the eye or a camera is—it will take in whatever light presents itself in the proper fashion. I see this again in the mystical tradition, this *via negativa* is really about developing what I call the muscles of receptivity. Our Western culture which is famous for its activism has very underdeveloped muscles of receptivity. We tend to fill this hole with junk because we're afraid of the dark or, if you will, afraid of nothingness.

Nothingness may be the bottom-line effort to name this unnamable mystery. No thingness. We have a need to experience nothingness and learn to be at home with it. Meditation practices are one such avenue, and suffering is another. I met someone who lost his life's work in the Oakland fire, for example; that's an experience of nothingness, being stripped down to the zero point. Maybe this concept of nothingness may shed some light on the physicists' effort to name the ninety-five percent of the Universe that is unnameable.

I want to come back again to this question of sleep. This is the way we all experience darkness every night—a letting go of all images and a kind of nothingness, although we experience it unconsciously. The point I want to make is this: no one has the faintest idea why we sleep or why any animal sleeps. Sleep is a profound mystery, and we take it for granted: of course we go to sleep, dogs and cats go to sleep, cows go to sleep, and so on.

I think I know why I sleep—because I'm tired, I want to regenerate myself.

Yes but bacteria, for example, don't go to sleep, and lower animals don't go to sleep. Sleep cuts in somewhere above the level of fish. Fish don't go to sleep in any normal sense of the word.

Maybe we sleep in order to dream.

Maybe we do, but the interesting thing is that it's not just we who sleep; it's practically all mammals.

But I think they all dream, don't you? I know my dog used to have nightmares and things.

Yes, they do. Insofar as we can tell from rapid eye movements (REM) or from measuring brain activities, dreams are widespread in the animal kingdom.

That's why I say that sleeping is not that good an analogy of going in the dark because we're always dreaming various things. We don't remember all our dreams but we remember a lot of them.

But the percentage of time that we're dreaming isn't that great. Periods of rapid eye movement, which are associated with dreaming, typically occur after intervals of about ninety minutes of dreamless sleep. The majority of the time we're not dreaming. If we're in a kind of nothingness, presumably there's nothing to differentiate my nothingness from your nothingness and maybe the nothingness of a dog or a cat or a snake—there's a loss of differentiation and boundaries.

If we're free from all thoughts and images, then there's no differentiation of my consciousness that distinguishes it from yours. There may be a linking up of all minds when they're asleep, a community of sleeping creatures. As you said, this darkness the mystics cultivate gives an extraordinary receptivity. Likewise, the darkness of our sleep may give us an extraordinary receptivity, and dreams may arise not only from our personal unconscious, or even from the collective unconscious, but from a much larger system which links us to nonhuman nature. It could be that there's a kind of nonlocality in that darkness, in which nothing's distinguished or separated from anything else. Perhaps the dark part of every light wave and every sleeping animal goes into this state that mystics talk about.

This does not trivialize what the mystics are doing. Mystics may be the people who have seen and understood that reality as we experience it rests upon darkness.

That really honors darkness. I love that image of things coming together in the dark. Because in the darkness unity, not differentiation, is what is experienced. In the light we can see so clearly how different we are in shape, form, and color, and every image is unique. That's why we get worn out at the end of the day—we're seeing all these differences. We need the darkness to reexperience the unity. As you say, the unity is primal; underneath all differentiation it's the unity of being that's most radical, most basic. Indeed, in the new creation story, it's that that came first. Togetherness came first and that's why all our return to darkness is truly a return to origin. It's a recovery of our original togetherness.

Again I'm reminded of that amazing passage in Eckhart when he's talking about the Godhead and he says, "When I return to the Godhead, the source, the core of all things, no one will have missed me," because there is such unity that "no one will ask me, 'where have you been, brother Eckhart, and what have you been doing?' "

In the light, we're judging one another all the time— "What are you doing, where have you been to legitimize your existence?" But in the mystery, as Eckhart sees it, there is no time and no light for judging. Things are so intimate that no one even notices our wandering out or our return.

Just one further point about sleep. One of the characteristic diseases of the modern world is insomnia. Sleeping pills are a major industry, and places like Los Angeles seem to be filled with insomniacs.

This may be related to your point about the modern world being saturated with light. We have lights blazing in our cities twenty-four hours a day. I've never heard of insomnia being a big problem in third world countries.

Maybe insomnia in Los Angeles and other cities of the overdeveloped world is the price we pay for the Enlightenment. We have light on demand, light twenty-four hours a day. It's as close as your television set: you can watch *The Tonight Show*, then old movies, and then it's time to go to work. Maybe this is really the Enlightenment coming to its logical conclusion. While, obviously, we can acknowledge the accomplishment of the Enlightenment in discovering a lot of facts and truths about nature, we're just beginning to realize the price we pay. Insomnia may well be only one price.

We invented this light machine, that's what television is, there's no question. That's why you can't avoid it. Your solution for your children is not to have one. If it's there, no matter how good the will and intelligence, if a light machine is on in the room, we're all going to be drawn to it, instead of paying attention to our stillness, or having intelligent conversation and family gathering, or whatever.

And it's a substitute for what was the primal light source for people, namely the fire, which is a focus.

We put a fireplace back in our living room in London, where we live in a Victorian house, and everyone who came into the room said, "Oh, it makes such a difference to the room, it gives it a focus." Because so many people said the same thing, I looked up focus *in the dictionary and found that it is the Latin word for* hearth. *So saying that the hearth gives the room a focus is a tautology.*

And hearth is made into heart, isn't it? Of course, a fire's so different from a television set. There are some passages in Eckhart where he writes about God crackling, that it's obvious he's at a fireplace and getting these images. You know a fire is going to crackle and it's going to sparkle, but you don't know where. It's so interesting, so unpredictable, there's so much wonder in watching a fire.

Yes, but a fire is something that most modern people hardly ever see. Whereas all traditional people have an intimate experience of fire every night, and every time they cook.

It wasn't just a pleasant distraction, either, it was for survival.

The evolution of human consciousness, most people seem to agree, has gone hand in hand with the use of fire. When people try to say what distinguishes people from other animals, the use of fire is often cited as one of the most important characteristics. This precedes the use of language prob-

ably by hundreds of thousands of years. The general
consensus at the moment, for what it's worth, is that lan-
guage may have originated between fifty and seventy thou-
sand years ago. Whereas fire-using has been around for at
least half a million years. Language may be a relatively re-
cent development.

Fire is one of the primary things in the evolution of hu-
manity. No other animal uses fire, and the use of fire is
ancient. It may be that the use of fire, and especially fire in
the darkness, has had a key role in the evolution of human
consciousness.

I love that line in John of the Cross about his escape in
the dark. It was a very political poem because he literally
did escape from his prison in the middle of the night when
there was no moon. That's why it was a "lucky night." He
says he had nothing to guide him "except the fire, the fire
inside."

That metaphor or archetype is of the fire inside the heart,
the heart as fire and being on fire, the burning heart, the
passion. We have the expression of the Holy Spirit as fire in
the Pentecost event.

I want to pay attention to an image you've brought up a
couple of times which really deserves attention in a discussion of darkness, and that is the cave, the return to the cave.
Francis of Assisi was always looking for good caves to pray
in. What is the cave? It's certainly, archetypally, a return to
the womb, a going into darkness and wetness and mystery
and unpredictability and lack of control. We don't know
what beasties are in this cave, and it takes courage to go in
and get out of the light. But also by doing that you're going
to encounter some side of yourself or some side of God that
is worth encountering.

I think the sweat lodge is a cave too. These are very

maternal images. Here's another spiritual path, to consciously seek out caves. At Chartres cathedral, one of my favorite parts is the old church down below, which has no windows and is utterly subterranean. It's a cave experience. You turn off all the lights and it's a cavern. Of course, the early church prayed in catacombs underground and it must have been very dark in there, because they were hiding from their enemies above.

I wonder about this return to prayer in the dark. You mentioned sensory deprivation tanks. Maybe to get people out of the light and all these causes of insomnia, to get them out of the Enlightenment fallout, the churches of our day should not be aboveground, they should be subterranean experiences. To lead people into the cave and there to do our prayer would be much more welcoming to silence. Going beyond the eyes. Rilke saying "the work of the eyes is done" means the work of the light is done and now it is time to go into heart work, which includes dark work.

You can't do heart work very well if you're all in the eyes or in the ears. I just wonder if the new architecture for a church should not be to invite people into subterranean caves and cellars where the heart is much more ready to experience the communion of one another: less judgment, less left-brain theology and more happenings for the heart.

The earliest paintings we know of, around thirty thousand years old, such as those at Lascaux in France, occur deep inside caves. They suggest that caves were used long ago for religious, initiatory, or ceremonial purposes. To get to them involves a long and hazardous journey. Presumably people would have had tapers or torches to light the way in. But why did they do it? It was obviously central to their practice in some way.

Recent research has shown that, in the places you find the pictures of ungulates, of hoofed animals of the kinds that they probably hunted, the acoustics are extraordinarily resonant. Whereas in places where there are pictures of catlike animals, there's usually very little resonance. The researchers who discovered this have speculated that in the places where there are pictures of animals that were hunted, the resonance would be associated either with chanting or with some kind of drumming or percussion. One suggestion is that the echoes might have represented the sound of the hooves of the animals running. So there'd be some kind of acoustical equivalent of the animals running, and the animals are indeed depicted in motion.

A movie theater.

Yes, and the images in the cave would presumably have to be illuminated by some kind of torch, giving flickering effects.

There may well also have been initiation rites there which involved long periods of darkness in the caves; as we discussed earlier, this would have resulted in vivid imagery.

It's interesting that this is the earliest thing we know about religion in Europe. Many thousands of years later, in ancient Greece, the Eleusinian mysteries were an initiation ritual in a cave, perhaps involving psychedelics. Certainly the initiates saw visions while they were in the cave.

Caves have also been used for religious purposes in many other cultures. The earliest Hindu temples that we know are carved out of solid rocks, and the earliest Buddhist temples, such as those at Ajanta and Ellora in central India, are cave temples. Later Indian temples were built out of stones but still involved the creation of a cavelike space for the holy of holies. The image of the god or goddess is in darkness, and to reveal the image you need a light. The Brahmin priest *lights a camphor lamp and circles the image with light, re-vealing the image in the darkness of the cave. Then he brings the lamp to you and holds it out so that you take the light with your hands and bring it to your eyes. The divine image is revealed by the light within the darkness of the cave.*

Likewise, in Europe many medieval churches and cathedrals have cavelike crypts below them. Crypt literally means hidden, as in cryptic. These crypts are usually the oldest part and in many cases the most mysterious and pow-erful part of many cathedrals, as in the case of Chartres.

Crypts are also associated with death because dead peo-ple, including saints, were buried under churches or cathedrals, and some crypts had recesses for storing bones. The association of the crypt with death and darkness is very clear.

Services are still held in some crypts, as at Canterbury Cathedral. But not in darkness, because they use books and keep the lights on.

Keep the lights off. I bet the service is no different from a service above; they don't take advantage. If we reinvented church architecture we'd build nothing but crypts. All new churches would be underground. But you would need some artists to reinvent ritual that really honors and even takes us back to those cave moments in some fashion, somehow hon-oring the darkness and putting people back into it. I think we are starved for darkness today. That's your point about insomnia—we're starved for darkness. Perhaps something religion could do for us would be to feed us some darkness.

This is a gift religion could give—and now at a time when it seems to have very few to give. Psychology has taken over its confession role and science has taken over its creation story role, so it's kind of running out of work. You used a

phrase that struck me when you spoke of "sensory deprivation." This has a lot to do with what I call the "mystical
tactics," whether it's celibacy, fasting, vegetarianism, or silent sitting. These tactics are recommended as one path into
our consciousness. Maybe this is an answer to why we sleep
at night: we need some sensory deprivation because we're
getting all this sensory input for two-thirds of the day. And
maybe that's the same reason Gaia sleeps at night—she too
needs some sensory deprivation to enjoy her own sunrise
the next day more fully. We have to be empty in order to be
full. That's the pattern, isn't it?

There's one story I want to tell that's very moving to me.
We've been talking about darkness and eyes. A friend of
mine, a student in our program at the institute, died of
AIDS a couple of years ago. He checked in to the hospital
and he died shortly thereafter. Everyone was surprised at
how quickly he died.

When I went to visit him and said goodbye, I was leaving
town for a day and I thought for sure I'd see him again.
There was no evidence that he was going to die that fast. As
it turned out, that was the last time I saw him. But I'll never
forget when I leaned over his bed just to say goodbye: his
eyes—he was an Irishman and had bright blue eyes—his
eyes were the blackest thing I've ever seen in my life and he
pulled me into his eyes like whirlpools, into black whirlpools. In retrospect I realized this was his goodbye.

This whole notion that our eyes can turn from a bright,
translucent blue into total blackness has really stayed with
me. They say that the eyes adorn the soul, that there is some
special connection between imagery, light, angels, photons,
and our eyes. The fact that his eyes had already made the
journey into the dark before his whole body did is very
interesting to me and I wonder if there's any lesson there.

Your experience raises the paradox of darkness and light in death. The image we usually have of death is darkness, blackness, and yet from near-death experiences many people describe finding themselves going through a tunnel and coming into the light, a light of great intensity. The Tibetan Book of the Dead speaks in a similar way of going into the light, so the darkness of death is associated with this journey into the light.

This interplay of darkness and light in death must have some relation to the mystical tradition. How do you see the connection?

Death appears dark to those who are living because there's a darkness to burying someone in the Earth and letting go of someone, the bereavement. Loss and grief are another journey into the darkness or into suffering. But— and this principle comes clearer to me because of your statements about the physics of light—light and darkness are moving along together. What appears very dark to us on Earth, at least to most people on Earth when someone they love dies, is in fact quite glorious and luminous and light-filled for those who are leaving Earth. That corresponds not only to the near-death experiences but to many religious stories, such as the risen Christ, and the sense of glory, *doxa*, radiance, that accompanies the stories of the resurrected Christ as told in the Gospels. The stories are about recovering our original radiance.

When Hildegard says, "Every creature is a glittering, glistening mirror of divinity" and "Every creature is a ray of God," she is naming this radiance, *doxa*, Shekhina, glory. We're dealing with polarities again of life and death, or death and rebirth, or dark and light. Death is an emptying, but it may also represent the polar opposite: a filling

with the deepest light and beauty of the one who under-
goes it.

Eckhart doesn't talk a lot about death, but he has this one
line which I think says it all: "Life dies but being goes on."

That is close to Aquinas's passion, and Walt Whitman's,
and Rilke's: existence is the miracle, existence is what
counts and what goes on. Again I go back to your model in
physics, in patterns of light and darkness. Presence and ab-
sence, life and death. After all, a being can't die without
being, otherwise it's nonbeing. I think it's rather rare to talk
about death as nonbeing; most traditions would envisage it
as another expression of being. You can have an argument
about reincarnation versus resurrection or something else,
but in some way I see death as a recycling process. In some
way we get recycled, it seems to me, into the pool of being.

*Most Christian teaching on the ultimate state of glory
suggests that the journey ends in light.*

Light has a last word, you might say.

Light has the last word. "Et lux perpetua luceat eis," *let
light perpetual shine upon them.*

Yet we have *"Requiescat in pace,"* may they rest in peace,
meaning may they sleep a really good long sleep. You would
presume that one sleeps best in the dark.

*It's a paradox because the two are coupled together, rest-
ing in peace and the shining of light perpetual.*

So there we have, maybe, the dance between returning to
God and returning to the Godhead.

This would be the ultimate polarity presumably. The state of glory would be the opposite of the darkness of the God-head.

Eckhart has this wonderful poem on repose. It's a sermon but it is a poem: "All things live for repose." This repose may be another way of speaking about eternal rest; it's another way of talking about the relationship to the Godhead as distinct from God. This is one more reason why church worship ought to go underground, since there's more repose there than all this business of getting the right pages and all that goes on in aboveground churches with the lights on.

I want to think of darkness as not just visual darkness but audio darkness, silence. For that kind of darkness I would propose that the Quakers have probably done the best job in the West of late. Honoring silence, being present to silence. That to me is part of darkness.

Eckhart too. When he says, "Let yourself go of all images," images are not just visual images but also audio images. That return in the dark to be emptied and cleansed is part of the darkness. When I imagine these cave services I don't mean memorizing banal prayers, I mean really entering into the darkness and being there together and seeing what prayers emerge from it all. Reenter ritual down there.

This would be a wonderful thing to do. Next time you're in Chartres or in England a place to allow this ritual to come through would be in one of these crypts in darkness or near-darkness, which isn't so difficult to do.

But there's another way. I would think as a naturalist you would be more interested in just getting a shovel and going into the Earth and digging out your own cave. Because the crypt has this mixed message: anyone can turn the lights on at any time or come charging down there leading us in some kind of banal prayer. Instead we can just go find a piece of Earth and start digging our own church basement, without the church. I also think catacombs are fascinating and full of potential.

Here's a story: Years ago when I was living in Chicago I was approached by some creative parents whose children were not having a good time, to put it mildly, at religious education classes in the church. These were fifth grade boys, about six of them. Their parents came to me and said, "Would you instruct our children in religion for us?" I agreed, and as part of my instruction I had these boys build catacombs in my basement. Every Sunday we'd build catacombs. There was a lot of paste all over the place and two-by-fours and chicken wire. When we built these catacombs we'd sit in there and tell stories. Of course, being fifth grade boys, they had to have skeletons, and we actually had a Mass in there as well.

Recently when I spoke at the Chicago Art Institute a twenty-eight-year-old man came up to me and said, "Do you remember me?" I said, "No." He said, "I was in your catacomb class when I was in fifth grade." "Oh," I said, "I remember that." He said, "What I do now for a living is I hunt stones and carve them. I've been to the Himalayas and been in an avalanche there, and into the wilds of Canada and everywhere I go I hunt stones." He handed me this beautiful stone on which he had carved all seven continents. "This is my work and it all began in the catacombs." I said, "It did? Why?" He said, "You handed out stones and told us to make up stories." I said, "I don't remember that." That was a wonderful encounter. I hadn't seen the man since he was just a little boy eighteen years before. I was taken aback that his vocation began in this make-believe catacomb.

Ten minutes later another fellow came up and said, "Do

.you remember me?" And I said, "No." He said, "I was in your catacomb class eighteen years ago and now I'm an artist." I said, "That's wonderful! You should go check in with that fellow." "Oh," he said, "I haven't seen him in eighteen years." Then he said, "My vocation began in the catacombs; in that class I was given permission to be imaginative."

All I can say is this: if this can happen with chicken wire and wallpaper paste in the basement of my apartment, what could happen if we took catacombs seriously again and started praying in these interesting cavelike places. Who knows what might happen. I hadn't heard from these fellows in eighteen years, and I had no idea that their vocation began in that simple occasion.

I think that's a great idea.

So in some ways theologically what I'm talking about is that it's time to start praying with the Godhead instead of just with God. The results with God are not really impressive since we keep turning out banal prayers. Is God banal or are we banal? Maybe we should begin at the beginning, which is darkness. Let go of all the prayer books and get into the dark.

Another way of talking about light and darkness is left and right brain. In a way I think that we've been saying that the right brain is not honored enough, certainly not in the modern and Enlightenment culture.

For Catholics, when liturgy was in Latin it was mysterious and entirely unknown to the left brain. So what really went on in worship was the smells—the incense and the wax candles—and the chants, and there really was so much more room for the right brain, for silence and darkness. There always is when you pray in an ancient archaic language, and surely this is one of the reasons why archaic languages are important in religions around the world. They invite the left brain to go to sleep.

What about ecumenism and darkness? In the dark you don't know if there are crosses on the wall, or Jewish stars, or a Buddhist statue. No one knows so everyone can be at peace. We won't be fighting one another there, we'll be praying together for a change. Sweat lodges are like that too. Once you're in there you don't know who's white and who's Indian and who's Christian or whatever—because it's dark. Togetherness can happen there.

CHAPTER SIX

MORPHIC RESONANCE
AND RITUAL

RUPERT

First I'm going to talk about the hypothesis of morphic resonance and then show how it may shed new light on rituals. To start with, I need to put this radical scientific hypothesis in its historical context.

In the background is the current clash between the two most fundamental models of reality in Western science. One is the model or paradigm of eternity, the idea that nothing really changes. The other is the paradigm of evolution, the idea that everything changes and develops in time.

The paradigm of eternity is rooted in the thought of ancient Greece and is in turn derived from the mystical insight of a timeless reality behind the changing world we experience. The ancient philosophers tried to find ways of putting this mystical insight into the form of intellectual theories. The Pythagorean school maintained that the eternal reality was mathematical. Numbers, and the ratios between numbers, have a curiously eternal quality, and they thought that this was the timeless basis of the world of change.

The materialists thought that matter was the eternal principle. Matter was made up of lots of little bits, the atoms, that never changed. They proposed that the world of change we experience comes about through the permutations and movements of the atoms, which themselves don't change.

Plato generalized the Pythagorean theory into his well-known philosophy of Forms: the world we experience is a reflection of changeless Forms or Ideas that transcend space and time. Everything in this world is a reflection of an eternal archetype. Each horse, for example, is a reflection of an eternal horse Idea.

This philosophy was Christianized by St. Augustine and others who said that the Platonic Forms are ideas in the mind of God. Platonism and Neo-Platonism had a profound influence on the thinkers of the Renaissance, particularly on Copernicus, Kepler, Galileo, Descartes, Newton, and the other founding fathers of modern science. They thought that through science they were discovering the true mathematical ideas about reality, laws in the mind of God. They thought that God was a mathematician, and mathematicians still like to think so.

When Newton combined this Platonic notion of eternal laws with the atomist notion of eternal bits of matter, he created a cosmic dualism on which mechanistic science has been based and is still based to this day: an eternal quantity of matter and energy, governed by eternal laws which are themselves nonmaterial and nonenergetic.

This paradigm has shaped the thinking of the physical sciences very deeply. Most scientists still take it for granted that Nature is governed by changeless laws. The idea that

*any experiment should be repeatable anywhere is a conse-
quence of this way of thinking.*

*By contrast, the evolutionary paradigm comes not from
the Greek part of our cultural heritage but from the Jewish
part. It is based on the metaphor of a journey, the prototype
being the journey of the chosen people out of Egypt through
the wilderness and to the Promised Land. This was extended to the idea of a journey in time of the Chosen People,
culminating in the coming of the Messiah. In the Christian
context, history was seen as leading towards the Second
Coming and the millennium.*

*This vision was secularized in the seventeenth century to
give the idea of progress through science and technology. A
millennial transformation would be achieved through con-
quering Nature. By the end of the eighteenth century the
philosophy of human progress was widespread. But Nature
was still thought to be static. It was only in the latter half of
the nineteenth century that Darwin made the idea of evolution scientifically credible. It was as late as 1966 that physics
finally adopted an evolutionary cosmology with the Big
Bang theory. Before that, most physicists still thought that
the Universe was eternal.*

*Now we have a radically evolutionary cosmology, and
this is at the root of the contemporary crisis in science, the
clash of paradigms. What about the eternal laws of Nature if
the whole of the Universe is evolutionary? If the Universe
evolves, why should not the laws of Nature evolve as well?
After all, human laws evolve, and the idea of natural laws is
based on an analogy with human laws. There seems to be no
good reason for assuming that all the laws of Nature were
established at the moment of the Big Bang, rather like a
cosmic Napoleonic code.*

The essence of the idea of morphic resonance is that the

regularities of Nature are more like habits than laws, that they're not fixed for all time from the beginning. They're habits which have grown up within Nature. Nature has a kind of inherent memory rather than an eternal mathematical mind. Salt crystallizes the way it does because salt has crystallized that way before so often. The more often it happens, the deeper the habit becomes until it behaves as if it's governed by eternal laws.

Each kind of thing has a collective memory of previous things of that kind. The basis of my hypothesis is that this memory depends on the process I call morphic resonance, the influence of like upon like through space and time. Similar patterns of activity or vibration pick up what's happened to similar patterns before.

This means, for example, that if you make a new chemical compound, the first time you make it the substance may be very hard to crystallize because there isn't a habit for that kind of crystal to form. But the more often you make it, the easier it should get all around the world. In fact, it is well known that the more often you make crystals the easier they crystallize all around the world. Chemists explain this away in terms of anecdotes which are part of the rich folklore of chemistry. The usual story is that this happens because fragments of previous crystals are carried from lab to lab on the beards of migrant chemists. If there haven't been any migrant scientists, it's assumed that microscopic dust particles of the previous crystals have been wafted around the world in the atmosphere.

Nobody has ever tested these conventional theories. I am suggesting that they can be put to the test by filtering the air in the lab and by excluding migrant chemists. I think that the crystals would still form more readily, even when these factors are excluded, because of morphic resonance. Thus one way of testing the hypothesis of morphic resonance is in *the realm of chemistry. Another way of testing it is in the development of unusual forms of plants and animals. (Details are given in my books* A New Science of Life *and* The Presence of the Past).

The hypothesis also applies in the realm of behavior. If rats in Sheffield learn a new trick, rats all around the world should be able to learn it quicker just because the rats have learned it there. The more that learn it there, the easier it should get everywhere else. There is already evidence from experiments with laboratory rats that this remarkable effect occurs.

In the human realm, the same principles should apply. It should be easier for people to learn what other people have already learned. The more who learn it, the easier it should get. Again, there is evidence that this actually happens.

This idea is already familiar to many people in the form of Jung's theory of the collective unconscious. Carl Jung, the Swiss psychologist, thought that all human beings draw upon a collective memory, which he called the collective unconscious. His evidence came from cultural parallels and from dreams that resembled myths in foreign cultures the dreamer had never heard of. He thought of these habitual patterns as archetypes in the collective unconscious. I think of them as morphic fields, which is the word I give to the patterns that are sustained by morphic resonance.

In nature and in human affairs, morphic resonance makes things happen in an increasingly habitual way through repetition. The key to morphic resonance is similarity, and its usual effect is to reinforce similarities as habits build up. Usually habits become increasingly unconscious. But sometimes, in the human realm, things are deliberately and consciously done the same way they were done before. This is especially true of rituals.

All human societies have rituals. Whatever the ritual,

whatever the society, whatever the context, rituals have a very strong conservative tendency. People believe they should do the ritual the way it has been done before. Why? They believe that by doing so they connect in some way with the people who've done it before. In tribal societies, through rituals the ancestors are believed to become present to those who are participating in the ritual now. Similar beliefs are found in all religions. The Jewish Passover feast is one example. The same ritual food, the same readings, the same remembering of the first Passover in Egypt are practiced year after year, over many generations.

In all traditions there tends to be a high degree of conservatism in ritual language. In India, Brahmins carry out their rituals in Sanskrit, because that is the old ritual language. The Roman church used to do them in Latin. The Russian church still does them in old Slavic. It is very common to find ancient languages conserved in ritual forms. Often people have forgotten the reasons for them, but they believe that the rituals should be done the traditional way if they are to work.

The purpose of ritual is to connect the present participants with the original event that the ritual commemorates and also to link them with all those who have participated in the ritual in the past. Ritual is something to do with crossing time, annihilating distance in time, bringing the past into the present. The Christian Holy Communion, for example, re-creates or connects participants with the original Last Supper and also connects them with those who have participated since; it brings them into a connection with what is called the Communion of Saints.

From a secular, rationalist point of view, none of this makes sense. Ritual is just one example of superstition to which human beings are unfortunately prone, until these irrational hangovers from the past can be eradicated by en-

lightened education. But rituals are remarkably persistent, even in secular societies. People seem to have a need for them, and when deprived of traditional ones, they re-create their own. Gangs, for example, often re-create initiation rituals, having been deprived of socially approved rites of passage.

From the point of view of morphic resonance, rituals make perfect sense. By consciously performing ritual acts in as similar a way as possible to the way they have been done before, the participants enter into morphic resonance with those who have carried out the ritual in the past. There is a collapse of time. There is an invisible presence of all those who have done the ritual before, a transtemporal ritual community.

MATTHEW

I think this is a rich concept, morphic resonance. Its implications for ritual makers at the end of the twentieth century is profound.

One definition of *liturgy* is *celebration as education.* When I was growing up we were told rather sheepishly that liturgy had something to do with education. Then no one ever mentioned it again. Now I'm grown up enough to see how limping our education is, how it often leaves us more wounded than enlightened. It's more important than ever to ask how we can revivify education, how it can be reinvented. I'm drawn back to this original statement that liturgy or ritual is a form of education.

How else do we really experience university, if university means a place to go to find your place in the Universe? It's rare to experience that in a university at this time in Western history. But you can experience it in ritual, because ritual draws on collective memories and myths and ancestral wisdom by way of awakening images in us by which we can connect to the whole. There's no celebration, it seems to me, without connecting to the whole, to the great picture.

The morphic field, the Communion of Saints, the ancestors and the future—all that power comes into a place when we are doing ritual well together. You might even say that many morphic fields are tapped into: the morphic fields of our ancestors and the morphic fields of the angels and the morphic fields of our youth and even of the future.

The education that takes place in ritual is about cosmology. As you said, the new cosmology represents a radical new view. How are we going to get this radical new view quickly into our bloodstream, into our dreams? It seems to me the shortest route has to be ritual. This ritual is how ancient peoples always told their young the great stories of the Universe. Ritual is how we learn the source of our existence and therefore the basis of our morality. Once we know where we are from, we can begin to agree on where we are going and how to get there—that is, morality.

Our new cosmology, with its new Creation story, brings back wisdom itself because wisdom comes from feeling connection to beginnings. If there's wisdom and a connection to the beginnings, then there's morality. Our new sense of values and morality is going to come this way. Certainly our experience of community finds its celebration, its naming, its high point, in ritual.

Experiencing an inner connectivity, we run with this sense of the Universe. A Creation story is ancient but has been retold by scientists in our time.

Then there is curiosity. St. Thomas Aquinas says that without curiosity there is no wisdom. Yet with our prefabricated and book-bound liturgical services in the West, where is there room or inspiration for curiosity? Seldom are surprises allowed to emerge in Western ritual today. Even the Spirit is locked into a box. So too are the angels and our bigger selves, our souls. Whenever I've dropped the line here or elsewhere about how boring worship is and how it's a sin to bore people and call it worship, I always get a good laugh. Everywhere in the world from New Zealand to Australia to the Americas, people recognize this about boredom and worship. What's the opposite of boredom? Well, one opposite is curiosity. Ritual somehow has to enliven our curiosity. Maybe that's what *interesting* means. If you attend native rituals, there are different rattles and different sounds and different drummers and different dresses and different smells — it's not uninteresting. To anyone, whether you're a baby or a teenager or an old person, it's always interesting, there is always something new. Like the Universe itself, healthy ritual is beautiful and interesting every direction one turns.

Another dimension to education and liturgy is *participation.* We learn by doing things, by doing things hands-on. Therefore, all healthy ritual is an invitation to doing things together, not being spectators watching someone else do something, but inviting people to truly participate with their own bodies, their own images, their own breath, and their own visions. This is one way that we awaken the morphic field and enter more fully into it.

A further dimension to ritual and liturgy is the theme of *remembrance.* This is the heart of all Jewish liturgy and ritual. But it's not just about remembering human events like Passover and the Exodus and human liberation. It is also about remembering the Creation events — the new moon, the equinox, the solstice, the seasons, and the sun, the birthdays and the death days of the galaxies and planets and stars.

Rituals and festivities are built around Creation's movements, because this too is a remembering. We've talked before about how mysticism is not taking for granted, in other words, not forgetting.

When Jesus at the Last Supper, as reported in the Christian Bible, says, "Do this in memory of me," he is simply uttering a thoroughly Jewish statement, that the essence of religion and the essence of ritual is healthy remembering.

I would rather talk about remembering and connecting than the word *conservative*, which you always use, Rupert. Because I feel that calling ritual conservative has such political overtones and it gives too much energy to those who are in fact turning ritual into a political conservative statement or into a nostalgia about their religious fantasy life when they were children. It feeds the religious immaturity which is all around us in our culture.

When I was living in Paris years ago, I was living with an Integrist family. The Integrists are fundamentalists, extreme right-wing Catholics, very violent people. I didn't know all this when I moved in with them, but one day the woman of the household said to me, "Since the Vatican Council I've lost my faith." And I said, "Well, what's your faith?" She said, "The pope and my catechism." So I said, "Well, you know the pope called the council and signed all the documents?" "No," she said. "Then my faith is just my catechism."

She just dropped the pope like that; he didn't have a chance. She was about sixty-five years old. So then I said, "I would love to see your catechism, would you please produce it?" "Oh," she said, "it went out of print in 1928."

That's what I mean. That's not the kind of religion I want to encourage when speaking about this dimension called ritual. I would rather use the word *connecting*. We connect to the past, of course, we bring the tradition along and we open up to it.

I also want to stress that I think ritual is a remembrance not just of the past but of the future. Aquinas has a beautiful poem about the Eucharist where he talks about tasting the eschatological food, the experiences of heaven and so forth. There's an eschatological dimension to ritual. *Eschatology* means the future and its coming to bear on the present. In ritual we're remembering what hasn't happened yet. We're projecting ourselves forward to experience the justice we haven't experienced yet, to experience the joy that isn't visible yet, to experience the celebration that hasn't broken through in our personal lives or in our social lives yet.

That's very important, that ritual is forward-looking as well as backward-looking. Eckhart says, "In the depths of the soul God creates the entire cosmos, past, present and future." If ritual and the entrance into the morphic fields is an entrance into our depths, then all time—past, present, *and* future—is going to come together there and some amazing things can happen as a result.

One question I have is this: Can computers, multimedia, techno-art call in the echoes and the voices of our ancestors and the angels in our time? How do they do it, or how do they not do it? Can the Christ who calls himself the light in all things arrive on beams of light, because in this century we have learned to harness beams of light? Can we take them back from the corporate makers of television? To me it comes down to a basic question—how do we tap into the morphic fields? How does ritual allow us to do this, and have new means been discovered in our time? The theology of light is so rich that when you tap into the technology of light in our generation, you are not just conserving but connecting to and bringing forward a powerful revolution.

At the same level, how does silence and how does darkness bring us into the morphic field, and how does it accomplish this in new rituals being made today? How is darkness a source? Darkness includes our own suffering. How can we create rituals that recycle suffering or, because suffering has a history, that connect us to the morphic field of all those who have suffered? This is the archetype of the Christ on the Cross. What tremendous energy there is there to recycle not only the life and the light but the death and the resurrection of the Christ and indeed of all suffering beings.

How does the relationship of personal ritual connect to morphic fields? Not just the collective community ritual, which is surely the high point of ritual, but those personal rituals that prepare us for the community ritual.

If the morphic field is a field, then maybe this makes sense of Jesus' analogy of being shepherds. We are shepherding the lamb in us, that cosmic Christ Child in all of us, in the fields where memories lie, the memories of our ancestors and our own best memories.

The physicist David Bohm says that matter is frozen light. I'm fascinated with that image, that matter is frozen light. Television, it seems to me, is not matter, not frozen light. Television, I would say, is fluid light. It's light flowing, light images flowing. One thing we've learned this century is how to render light more fluid and more flowing.

Mechthild of Magdeburg, who was a wonderful Creation mystic and activist in Germany in the thirteenth century, was involved in the women's movement the Beguines. She talked about "the flowing light of the Godhead," saying that we should "lie down in the flowing light of the Godhead." How can new rituals get us to dance in this light as well as in the darkness and to lie down in the light and thereby imbibe the fire of the Godhead? So those are a few thoughts.

You're right about the unfortunate implications of the word conservatism. *The words* connection, remembering, and tradition *better express what I was trying to* convey.

In this context it is interesting that people who are engaged in ritual reform usually claim that what they're doing is not a true innovation but rather a return to some more ancient form. For example, in the liturgical reforms since the Second Vatican Council and in the Anglican church, when people objected to changing the traditional forms, the reformers said that whereas the familiar rituals had been going on for a few hundred years, the reforms were returning to the original forms of the early church. They scored a lot of points by claiming that they were being more traditional than the traditionalists.

The trouble is that we know very little about the deep roots of rituals. We know quite a lot about human history for the last 2,500 years. The great pyramids were built about 4,500 years ago. We've got cave paintings almost 30,000 years old in France, made by people who went deep into dark caves and saw these paintings by the light of flickering torches. You could even say that they were a precursor of television or cinema, flickering images in the dark in the cave.

But before that we have human beings going back a million years or more. We know almost nothing about what they were doing. We know they made stone tools and they had fire. Sitting around a fire at night must have been an utterly primal experience, and the flickering flames must have conditioned the whole evolution of human consciousness very deeply. But we know next to nothing about their rituals, although we can guess that they were related to shamanic practices and had fire as a central image.

For me it's an important thing to find ancient roots before I can take any new ritual practices seriously, and I'm trying to do that now. Electronic media, including TV screens, could, through their flickering, liquid light, tap into very ancient ritual morphic fields, while being thoroughly modern at the same time. This thought helps me to see TV in a positive way, because I am not very keen on it, and the idea of sacralizing TV is for me a new idea. I have until now seen TV as a godforsaken, secular medium.

One of the things that impressed me when I was living in India was the way Indians incorporate the new into the old. This was brought home to me vividly when I went to work in my laboratory on the festival of Durga Puja. On this day of sacrifices to the goddess Durga, people ask for her blessings on the instruments of their trade, scribes their pens, artisans their tools, bus drivers their buses, and so on. When I arrived at the research institute, I found to my astonishment the laboratory decked with banana leaves, the drying ovens and weighing machines garlanded with jasmine flowers, and limes under the wheels of our pickup truck ready to be crushed by the truck driving slowly forward.

In the mainframe computer room the terminals were garlanded, incense sticks were burning, and in front of the computer was a block of stone and on it was a coconut, which the American head of the computer services section was asked to break as a ritual sacrifice, to ensure the blessings of the goddess on the computer for the coming year. I had never before thought that modern technology could be brought within the purview of ritual. It was a very surprising revelation.

If we are to transform this very secular world in which we live, then ritual forms must somehow embrace and work through the technology. There's nothing new about light and there's nothing new about electricity. These are things that are fundamental to the whole structure of the cosmos from the time of the primal creative moment, the Big Bang.

We can find incredibly ancient roots for all the things that

exist in modern technology. The way they're used is new,
but the roots are very old. By thinking in those terms one
can see how the new can be incorporated into ritual.

Well, I agree, and again the word that you used a couple
of times is *tradition.*

When it comes to morphic resonance and technology, I
met a woman recently in her thirties and she was saying that
she is still intimidated by computers but that her fourteen-
year-old daughter is totally unintimidated. She'll sit down
and do anything. To me that confirms your morphic
resonance idea. This younger generation, being raised on
computers, is thoroughly at home with them.

Your beautiful story about technology being blessed in
the Indian scene seems very appropriate. Why wouldn't we
do such blessings for our computers and media and micro-
phones? If they are going to be tools of our liturgy, then we
have perhaps a responsibility to embrace them in ritual itself, make them really part of our ritual.

I'd like to pick up on the word *habit,* because when I hear
that word I always think of Aquinas defining virtue as a
habit. It seems to me that ritual is a virtue about virtue, that
it brings together habits of all sorts and tries to kind of
recycle them or order them, make them harmonious within
the context of the whole community.

But what about the dialectic between tradition, which is
something we're bringing from the past, and the moment of
creativity where new habits come forward. What do you
find as a scientist is the matrix in which new habits burst
forth?

If you look at the context for creativity in the biological
sense in evolution, new habits arise because old habits are
blocked. Most creatures are creatures of habit, including us.

And if we're able to go on with our old habits we usually do; this is the line of least resistance. Habits are very difficult to break even if you want to break them. Usually you're not even conscious of them. The point about habits is that they become unconscious. This is convenient because we can usually think about only one thing at a time; indeed, it is necessary that most of the time we are on auto pilot for most of our activities.

In general, habits are broken only by something that forces them to break. The dinosaur habit was broken by some enormous disaster that caused most of the species on the face of the Earth to become extinct about sixty-five million years ago. Nothing larger than a chicken walked away. Then there was a rapid phase of mammalian evolution over the next five million years, which is a short time in geological history. Most of the types of mammals we have today had evolved in this creative period.

Evolution involves an interplay between habit and creativity. Creativity can give rise to new habits. But in the biological, cultural, and religious realms, innovations are subject to a process of natural selection. Most innovations do not survive. But some are more successful than others and are repeated, and through being repeated they become increasingly habitual; their morphic field becomes stronger.

In the human realm, too, changes in habits usually come about because the old habits won't or can't work. Most of us don't willingly give up our old habits. We're usually forced to give them up because of a crisis, conflict, or disaster. This breaking of our old patterns can then open up a space for creativity.

In the context of liturgy, old forms can obviously become habitual and have a numbing effect. Here too the need for creativity comes about through the old habits not working. No one is more emphatic than yourself in pointing out that the old habits of church rituals are not working very well. If the old forms were working superbly well, there wouldn't be much need for liturgical innovation. But clearly times have changed.

But we should recognize that being run by habit is the norm in human affairs. Many traditional human cultures have remained stable for thousands of years, with few innovations. We now think rapid change is both desirable and necessary, but this is abnormal, even pathological, in terms of human history.

The same is true in biology. A typical species appears relatively rapidly, lives more or less unchanged for millions of years, and then goes extinct. In the case of living fossils, plants or animals go on virtually the same generation after generation for a hundred million years or more.

The cockroach is two hundred million years old, and it hasn't changed.

It's adapted its behavior, though, very successfully, to living inside New York apartment buildings.

In our new era, the postmodern era, what do you think about the role of the scientist in liturgy?

Most scientists consider that their role in liturgy is zero, because they have little or no interest in religion. Most of my colleagues are steeped in secular humanism and antireligious rhetoric. There isn't a big discussion of these issues among scientists, even among those who are interested in religion.

Perhaps the most direct role that science and scientists can play is through the amazing new visions of Nature that science has given us—the discovery of billions of galaxies, the expansion of the cosmos, the modern Creation story from the Big Bang onward. This is something that Brian Swimme and Thomas Berry have explored, but usually scientific discoveries about Nature are outside the purview of religion, partly for historical reasons. The deal done between science and religion in the seventeenth century was that science would take the whole of Nature, including the whole of the heavens, all forms of life, and the human body, while religion would take the human soul and concern itself with questions of redemption and morality.

Religion got the worst part of the bargain, and science ever since has proceeded as if the whole of Nature is entirely secular, devoid of any spiritual content, meaning, or interest. We need to recover a sense of the deeper meaning and spiritual dimensions of Nature in the light of astonishing discoveries we have about cosmic evolution, about biological evolution, about the levels and layers of structure that we find in cells, tissues, organisms, and atoms. The big picture that science has revealed is amazing and it's largely untouched by the religious imagination. I think this is an incredible area of opportunity.

Second, in addition to the new picture of Nature that science has given us, we have the new technologies. As we have already discussed, these are entirely secular in their conception, manufacture, and use. The challenge here is to find a way to discover their spiritual dimensions. So far, most have been kept out of traditional forms of worship as being unwelcome intrusions from the modern secular world.

Third, I think it's important for scientists themselves to recognize that what they are doing has a ritual dimension. Right from the beginning of science as we know it, in the early seventeenth century, Sir Francis Bacon envisaged scientists as a new priesthood with the whole of Nature as their sphere of activity, conferring great benefits to humans *through technology. In the nineteenth century this process was taken much further under the leadership of people like T. H. Huxley, who strove to professionalize science, squeezing out amateurs and natural historians, to create a secular priesthood in opposition to traditional religion. Most scientists think that they've won this battle with religion. The modern world is in the grip of science. Every country in the world has the same kind of science, the same kind of technology, but there are lots of different religions. Whenever science and technology take over, religions will sooner or later fade away. That's the standard model.*

But now this secular priesthood is in a state of crisis. It is fast losing its popular appeal. It doesn't appeal to the imagination, it doesn't have any public rituals, it has very little to fire a feeling of participation. There is a growing sense of public alienation from science, which has meant scientists are now demoralized to an unprecedented extent.

We have to recover a sense of the ritual aspects of science itself. Some of them are very unpleasant—animal sacrifice, for example. Millions of animals a year are, as we say, "sacrificed on the altar of science." I also think it helps to recognize that scientific experiments are a modern form of consulting oracles. In a real experiment you don't know what's going to happen. You do research only in areas that are undecided or unknown. If you know what's going to happen, you don't bother to do the experiment. In experiments you're always dealing with the undecided, you are on the very edge of what's knowable. This is the traditional realm of oracles. Some people peered into entrails of animals, others listened to the sound of the wind in oak trees, others went into trance as in Delphi and gave Delphic pronouncements.

Traditional oracles had to be interpreted by priests or soothsayers. Likewise, in most scientific experiments, the data you get are often ambiguous and confusing and they have to be interpreted. Scientists are like priests in an interesting sense insofar as they're reading the oracles of Nature. It would do science a world of good to recover a sense of ritual within its own activities. But this is far from anything I've ever heard discussed among scientists.

I once met a physicist at the University of Pennsylvania who was in his sixties. A group of us were in a discussion and he asked me a very complicated question about religion and science. My response was, "You know, most scientists I've met in the last fifteen years have told me they got into science because of a mystical experience they had when they were children, looking at the stars or worms or something else." His face became that of a little boy; he moved from being a sixty-five-year-old to an eight-year-old. He said, "That's it! That's why I became a scientist!" Then his mood totally changed, he became childlike, and there was awe in the room and I didn't have to answer his complicated question.

The real issue of how scientists join in ritual has to do with their personal story. What drew you into biology? What age was it when you knew you wanted to be a biologist and was it a mystical experience, an experience of wonder? You used that word *amazement*, Rupert. Aquinas said that the philosopher and the poet both begin with wonder. Well, I would say that the philosopher, the poet, *and the scientist* all begin with wonder. But scientists lose the wonder like theologians and priests and everyone else, because they go through academia where it's beaten out of you and you're rewarded for many other things, but not for wonder.

That's the common ground, this sense of wonder. Any scientist who still has it is welcome at a liturgy and those who don't ought to get recycled and confess their sins. They lost wonder somewhere along the way. Perhaps remembering wonder is the meaning of ritual as memory.

Did your vocation begin way back when, with wonder?

Oh yes. In retrospect I can see it. It took me a long time to remember, because in science people rarely talk about how they got into it in the first place. In my case it was an experience when I was about five. My family were willow-growers in Nottinghamshire. Near the farmhouse I saw a row of willow trees with rusty wire between them, and I asked why the wire was there. They told me that it was a fence that came to life. Seeing how those seemingly dead stakes could come to life gave me a tremendous sense of the power of regeneration in living things.

At Cambridge I did a Ph.D. on the regeneration of stem cuttings from a variety of plants, including willows. In India, I developed a new agricultural system based on cutting back pigeon tree plants and letting them regenerate new shoots. In retrospect most of my scientific career, even my recent book The Rebirth of Nature, *has a general affinity to this primal epiphany.*

This realization made me ask scientific colleagues why they did what they did. In almost every case they at first had no answer, and then later they came back to some primal moment of wonder which had been completely forgotten, or at least deeply buried.

Many scientists do indeed go into it out of a sense of wonder, but then there are years of training. You learn, as I learned, that it's no use wondering what animals are thinking or feeling, or trying to empathize with them. You have got to cut them up and find out where the vagus nerve runs and that kind of thing. There is a process of emotional distancing, especially in biology. First you cut up dead animals and then you move on to cutting up living ones in vivisec-

tion experiments. And the natural repulsion that you have from doing this is something you have to harden yourself against. That is part of the initiation ritual of becoming a biologist.

By the time you're ordained, when you get your Ph.D., you've gone through a training that has extinguished these emotional responses, at least when you're in the laboratory. This has meant that a lot of the primary inspiration which led people into science is suppressed and ceases to be accessible. Then the scientific profession becomes driven by things like personal ambition, competition to get the results out first, winning bigger grants, empire building, and that kind of thing—all the familiar vices that are not just vices of science, but vices of so many human institutions and professions.

In many ways I see ritual leading the way for invigorating all our professions. What you said about scientists losing their sense of wonder and selling out could really apply to all our professions—lawyers and politicians and business-people and priests and educators. This sense of wonder is missing everywhere.

How is it going to come back? Primarily it is ritual that is going to bring it back. The implications for the other professions of this blockage or this breakdown, this failure of ritual in its current form, are tremendous. We need a reinvigoration of all our work worlds, bringing the new cosmology and spirituality into our work. How can that happen except through the storytelling that ritual is capable of? Certainly there's no swift way of doing it except that way.

Then there is the ecological crisis. The failure of our ecosystems is obviously due to habits—you know, this is part of morphic resonance that we don't talk enough about, Rupert. Not only do good habits get easier, but bad habits do too.

Consumerism is much more advanced today than it was one hundred years ago. So is what we take for granted—you've got to have a refrigerator, you've got to have a car, you've got to have air-conditioning, and many other things that our grandparents would never have thought were necessary for surviving on this planet.

Your theory is extremely important for that too, for shedding light on addiction and bad habits, and hopefully moving beyond them.

As you point out, morphic resonance would sustain bad habits as well as good ones; it would also help them to spread. I'm saying that habits are part of the nature of Nature, the nature of society and of human nature.

Bad habits can be a problem in religion as in everything else. Any habit, even a good habit, tends to unconsciousness. When rituals become entirely habitual, they become boring, lulling people into a kind of stupor, their minds wandering. But at the creative moments in religion, the habits are not yet established and do not have this dulling effect. The creative principle by definition is extremely strong at those originating moments.

The use of ritual in a proper way can enable people to resonate with those original creative moments and therefore connect with that original time of insight, making it present so that there can be a continual renewal of that creative potential. That is their aim, I suppose. The trouble is that the ritual forms can be cloned, but not the openness and inspiration. This is probably true in many other contexts and institutions, not just in religion.

Definitely. That original moment is a very important one. But it's inevitable, it's part of the law of evolution, isn't it, that you can't freeze the charismatic moment. Isn't that the same in relationships? Marriage and other relationships too? Or children. I've heard my parents say how cute we were as babies in a nostalgic way, you know.

But that's it, the temptation to freeze novelty and newness instead of keep inventing newness. And this is where the mystics are so powerful—Eckhart saying, God is *novissimus*, the newest thing in the Universe. That's why the Bible begins with the words "In the beginning." God is always "in the beginning." And Jesus, after being illuminated as the cosmic Christ at the mountaintop of the Transfiguration, advised his three friends to tell no one and to return to the city. No freezing of the breakthrough moment, there.

Our mystical tradition invites us to constantly be recovering this sense of newness, so every day is a new day for all of us. In this way the child in the adult does not dry up and become a cockroach content to live out of habit over two hundred million years. This is my understanding of how our species is unique. Compare a young chimpanzee to a young human. As a baby, a chimpanzee can outplay the baby human; but as it grows older it ceases to play. Humans, on the other hand, have the capacity to play even as adults. That to me is mysticism. What you're saying is really important. Can we give birth to institutions or traditions so that they remain young? And if they are young, then they are, in Eckhart's words, in God and God-like.

But if our traditions become institutionalized and fossilized and old in every sense of the word, then they have to yield to the new forces of spirit. Eckhart says, "The first gift of the spirit is newness." The Spirit is biased—and I think evolution really supports this—Spirit is biased in favor of the new.

CHAPTER SEVEN

REVITALIZING EDUCATION

MATTHEW

I've been struck by Thomas Kuhn's statement in his classic work on paradigm shifts that, in a time of paradigm shift, education is all-important. I think that's true. The only way out of human collapse and catastrophe is human imagination, mind, and creativity; we have to get our wits about us. We have to educate smartly.

If education is all-important, the implicit question that always comes up for me is: What education? What model of education? Education is also part of the paradigm shift. We can't solve the paradigm shift with the former paradigm or models of education. And if you look at the stagnation of our work world and its professions, whether you're looking at health care providers or priests and ministers or artists or politicians or businesspeople or economists, all along the line you realize they have something in common: they all went to school.

They all went through the same system. Maybe, just maybe, something is amiss with that system. If you can get to that issue, then there is some possibility of a paradigm shift. A priest friend of mine was hired at a medical school in New York City a few years ago. His job was to live in the dormitories to prevent suicides. The previous year four medical students had committed suicide and numerous others had attempted suicide.

That's rather interesting, that the profession which draws on the mythology of healing is killing its people in the process of training them. Then you have to ask how many others didn't commit suicide but are doing it in other ways. For example, through workaholism or through avarice. If we want to deal with the explosive health costs in the West we're going to have to start talking about the model of education in our medical schools and the kind of doctors it is destined to graduate. Does a soul become avaricious because it has missed the spiritual side of education? In all world traditions, healing has been considered an art of the spirit. The words *salvation* and *well-being* are the same words all around the world.

When we're dealing with health care, we're dealing with an ancient spiritual practice that our culture has managed to secularize. One result is that it's way out of line financially, and another is that it's not doing its job.

A study was done recently on doctors in America. They asked ordinary Americans how much money they thought doctors made and how much money they should make. The poll found that most Americans think doctors make $125,000 a year. The fact is the average doctor makes $250,000 a year, and the average American said they should make $100,000. They also asked how much money they thought drug corporation presidents made. (One of them in 1994 made $13 million.) Needless to say, the American public underestimated by about $12,850,000 how much presidents of drug companies make.

This is traceable to the educational system. Where else would it come from? Other cultures in the past have not had this bloated problem of avarice built into their professions. This is just one example of the doctors; lawyers would have their stories to tell.

We've underestimated how powerful education is, how easily it can distort an entire civilization. Why? Because it is dealing with the most powerful being on this planet, which is the human mind. You would think that we would be more critical, that we would stand back from education and say, Are we training the whole mind or are we just training this part? How would you go about training the whole mind?

The mind and body together.

Exactly. Mind, heart, the whole person. I would like to say as a blanket statement that our professions are dying essentially of avarice. That is why the culture is dying, because those who claim the spiritual work of training the mind, called educators, or the spiritual work of healing the body, called doctors, or the spiritual work of defending the poor, called lawyers, these people aren't doing their job. They co-opted the spiritual titles but they're doing other things with their power.

Now Aquinas analyzes avarice this way: it's not materialism, it's a sin of the Spirit. He says, "Avarice tends to infinity and knows no limit." So what is infinity? It's Spirit; Spirit is what is infinite. What we're dealing with here is a spiritual issue. People are looking for the Spirit and they have been taught to find it in the power of money, fame, prestige, political influence or what have you.

The question then becomes: How do you heal this avarice? It is by offering authentic experiences of the Spirit or of the infinite because avarice is built into every addiction. If you are an alcoholic, you never have enough alcohol until your body stops and you die. How do we cure this? We all do experience the infinite or the Spirit authentically, in three ways. First in the mind. The human mind can potentially know all things, as Aquinas says. Therefore, training the mind will give people real, satisfying experiences. The unsatisfying ones called avarice are infinite in their demands but they cannot satisfy.

Second, we experience the infinite in the human heart, which never loves too much. The human heart is infinite in its capacity for love but it depends on the mind to feed it daily with lovable objects. If the mind stops or closes up, then the heart freezes too because it's not being offered new objects of delight on a regular basis, to praise.

Third, Aquinas says that the human hands connected to human imagination produce an infinite variety of artifacts: creativity. From the first cave paintings to today no two painters have created the same painting; no two musicians have composed the same song or dancers the same dance or potters made the same pot. So here is a model for education that will combat avarice and offer us a whole different direction. And that would be developing the mind in all its capacities, right and left hemispheres and therefore the body and the heart. Developing the heart means body work, and developing imagination means art or creativity.

I don't think that's complicated. Schools at all levels of society need to follow this model since the ones they continue using are dangerous to the planet and to the rest of us.

RUPERT

What you are saying, it seems to me, is that we've got a thoroughly secularized education system. It teaches techniques that are supposed to be morally neutral. For generations, scientists have pretended that science is morally neutral; for example, there's nothing good or bad about making an atom bomb. If politicians choose to use it in a certain way, it's their fault, not the scientists'.

Well, that argument is utterly specious and has been most obviously shown to be so through the Star Wars program where scientists took the initiative in getting many billions of dollars to feed the budgets of laboratories that work on advanced weapons research. It was entirely initiated by scientists. These standard arguments about the neutrality of science are not valid. Nevertheless, scientists have pretended that science is purely objective, morally neutral, and that morality's something else.

Sir Francis Bacon first made this distinction between facts and values. Facts are the realm of science, and he equated them to the pristine knowledge of nature before the fall, when Adam named the animals. This happened before he ate the fruit of the tree of knowledge of good and evil, when morality came in. This was Bacon's principal argument for the validity of science, a theological argument. Ever since, science has rested on this distinction. The God-given ability of man to name and know the Creation comes first and is blameless. Morality comes later. The realm of morality is the realm of religion. The knowledge of Nature is the realm of science.

That's a pretty good deal you've cut out for yourselves. Does that make scientists less than human or more than human?

Just in a paradisiacal state of innocence that's only tainted later by people like politicians and priests. The innocent knowledge of Nature was the ideal. There's something to be said for it as an ideal in the naturalistic study of Nature. But of course Bacon took science to mean not just natural history, but the study of Nature for the sake of manipulating, controlling, and exploiting it.

This ambiguity was built in, and education in science has been a matter of learning facts, techniques, and so on as if they are all morally neutral. For example, education in medicine involves learning the names of bones and muscles, the kinds of germs that cause disease, the kinds of drugs that kill germs, the way the heart beats, the physiology of the nervous system, and so forth.

Because science has such high prestige, it then becomes a model for other forms of education, seen as secular and morally neutral, even in literary studies. You learn facts and techniques of literary criticism or art history. It's not a question of whether some kinds of art are moral or immoral, inspiring or degrading, it's a question of which painter influenced whom, and what kind of technique they used for their brush strokes, and how patrons affected them, who owned which pictures, and so on.

This secularized education system takes morality or values out of it, and they are relegated to a small part of the curriculum in religious studies or religious education, which in practice, at least in Britain, usually becomes wishy-washy comparative religion. The basic message it imparts is not a celebration of the richness of all different religious traditions, but the fact that any given religion is a culturally determined set of manmade myths and stories. They are all different and have no kind of objective validity, unlike science where there's only one kind. Science has become, in essence, a monotheistic system. There's only one kind of

*science worldwide whereas there are many kinds of culture
and religion, reflecting the foibles of human nature, of human subjectivity. Science is supposed to represent the true
objective knowledge of Nature.*

*This separation of facts and values leads to many prob-
lems. Right-wing politicians, in an incoherent way, are often
decrying modern education for the way it fails to inculcate
morality, blaming it for problems of crime, alienated youth,
and so forth. But I think, as you indicated, there's a much
deeper problem here. Not only is there no moral content,
but the spiritual side, the emotional side, and the bodily side
are left out as well, except in the small sector called physical
education, which often fades out by the time students go to
university.*

Much of it is about competition anyway rather than development of the whole person.

*There's this contraction of education to a fairly limited
area of facts and techniques, combined at the more advanced level with initiation into a professional league or
guild, almost inevitably dominated by an established paradigm. As Kuhn said, each profession has a collectively
shared paradigm. I would see it as a morphic field, a habit of
thinking.*

*There is nothing in all this to resist pride, greed, and
envy, which are deadly sins. Indeed pride, greed, and envy
are the engines that drive our modern economy.*

We're educating people for it very effectively.

*Yes. During the Thatcher era in Britain, the things that
were praised as making economic progress happen were
competition, the desire to get more and to get on. Of course*

the desire to shine over others can easily lead to pride or is motivated by pride.

I like the word *arrogance.* I think it's a better word than *pride* and that it says more to us today.

Yes, arrogance. Of course, education leads to the arrogant idea that if you've got a higher level of education you're better than someone who hasn't.

And you speak your own language.

You have a special language that other people can't understand, like medical jargon or the jargon in any profession. And this is something we've exported to the third world. Not that people need lessons in arrogance anywhere, but in India it's very striking. When a young man has a B.A. he can command a higher dowry with a potential bride; an M.D. commands a yet higher dowry, and so on. Even a "B.A. failed" counts for something. And then you can buy bogus B.A. and M.A. certificates on the streets of Bombay near the university for a few dollars.

Arrogance is one of the things that education certainly doesn't counteract, it actually favors. Greed dominates some professions, like law. Fortunately, there are other professions where it is not the reigning principle, like schoolteaching and nursing, professions which contain a large proportion of women.

Envy is the very basis of consumer society. Billions of dollars a year are spent on advertisements to excite and fuel envy. Almost every culture is susceptible to this. Introduce Western consumerism to almost any culture in the world and within one generation they're really into it. They want McDonald's, videos, TV, and so on.

Resentment, I think, is kind of a subcategory of envy.

But how could one have an educational system that deals with all aspects of the personality, not just various kinds of more or less detached mental activity? How can one change an educational system that, at any given time, is going to be imparting the reigning orthodoxy.

It's ideology, it's politics.

In a tribal culture, of course, this means the children are educated, or rather initiated, into the skills and customs of the society; in a more or less stable society, it is necessary that children be brought up and integrated into the society as it is, maintaining its own particular social, cultural, and religious system.

But we've got an unstable society set up in such a way that competition, greed, envy, and pride are regarded as good, or at least as necessary evils, because they propel economic growth. Having lived in third world countries, like Malaysia and India, I've seen villages where people were still living in the older, traditional way, with its built-in social and religious constraints on envy, greed, and pride. But when the educational system comes in, modern attitudes soon begin to make an impact. One of the first things it does is to raise young people's expectations, making them want more, making them more ambitious. Then they can be motivated to do all the things that developing economies need, such as become people who are prepared to go and live in smelly, crowded places, work long hours subject to an imposed discipline, turn up at regular times in the morning for work, and all that kind of thing. Schools themselves epitomize this disciplining process, subordinating both spontaneity and local tradition into a standardized pattern, laid down

*in ministries of education, advised by foreign experts from
UNESCO, and subject to the rule of the clock.*

*When I was staying in a Malay village, a kampong, there
was an old man who lived in a neighboring hut who was
extremely amiable, usually smiling, always with time for a
chat. One day I and my village hosts met him as he was
setting off on his bicycle to the local market to sell a few
bunches of rambutans, a delicious fruit in the litchi family.
In conversation he said it was an unusually good year for his
trees, with a good crop, at a time when the market prices for
rambutans were unusually high. He explained he was happy
because he needn't take so many to market; he could get
enough money by doing less work. A semieducated young
man pointed out to him in a rather superior way that he
could get much more money than usual if he took more to
market. But he replied, "Why do I need much more money?
I've got enough selling these." For him, his good fortune
with the rambutans was not seen as an opportunity to make
more money, but to do less work and to enjoy life more. He
was thinking qualitatively; the young man was thinking
quantitatively, in the modern style. The idea of "enough"
was the mark of the uneducated simpleton.*

*The modern mentality is that more means better; quantity
prevails over quality. That's the culture we live in, and I
suppose the educational system necessarily reflects it. But in
trying to imagine a different kind of education, how can new
paradigms be explored when there's a reigning paradigm in
force? The only answer seems to be to set up an alternative
model. It seems to me that we already have an alternative
model of education which is better than most of what happens in schools and universities. And that's in workshops.
You can go to workshops on many different subjects and to
learn many different skills, but the style of teaching is very
different from that in institutional education.*

For example, my wife, Jill Purce, gives chanting work-shops, and the people who come do so because they really want to chant. They want to learn by experience, not just intellectually. Most people who go to workshops get really fed up if someone just lectures at them. There's an implied participatory element in workshops, an experimental element, and an attention to group dynamics that is absent from most classrooms and lecture halls.

Workshops are not perfect, but in them I think we already have, up and running, an alternative kind of education. But the workshop type of education is functioning mainly for the middle-aged; in most parts of the Western world it has little contact with people in their teens or twenties.

Or the poor; it's a middle-class phenomenon.

It's a middle-class, middle-aged phenomenon. But this is one possible model of a different way of doing things.

Another possible breakthrough is in terms of some of the new technology. Younger people with computers are educating themselves through dialogues and discussions on the Internet. CD-ROMs are a fascinating way to learn also. This technology holds the potential to put some of the great thinkers on a disc and let people play with their ideas. The question becomes: Can young people educate themselves now that we have so much technology?

One thing that is clear is that young people like to play with the technology. They're good at it; they learn it very fast. It's like a morphic field; there's a memory developing and it gets easier and easier with each generation. And do we really need all this paraphernalia that goes into school systems? It's terribly expensive, since essentially learning is something that our minds are yearning to do anyway. For me it's a question of ideology. It's the original sin versus original blessing mentality. If you think you have to force learning into people, then perhaps you need these systems which in many ways parallel prison institutions. You need wardens and superintendents and all that.

But if you believe that the human mind is original grace or original blessing, that the mind yearns to know and will do so if it's led to the right waters, then it seems to me we could really simplify the whole process that we're calling education. Part of it I think is peer group work, getting people to educate each other. That's a big part of workshops too. It's what goes on after the sessions and between the sessions, dialogues with people.

And certainly the model that we've been putting out for eighteen years at the Institute in Culture and Creation Spirituality is a conscious effort in this direction. I too have always believed that these retreat and workshop models are truly an alternative. But we also try to work within the system we have so that the accreditation is still there. Workshops don't give accreditation, they don't give real degrees. And degrees are important to people for financial and security reasons. It is important to work within the system but offer what I would call curved instead of square models of education.

The kind of work we do with art as meditation, with body work, rituals, along with intellectual work, combine right and left brain training. It's very effective. There is a lot of energy. It's a transformative process for people, and I would like to see this shared at all levels. I think often of how our children, the youngest children, still get crayons and are encouraged to draw and to use their imagery faculties somewhat. But it's right in their adolescence when they need this experience most that we take it away and make them so serious about getting into college and about doing only the left brain work.

It would be folly of us at this time not to take a hard look at aboriginal peoples and ask how they educated their young. Their culture's lasted tens of thousands of years. Ours was ready to blow up the world after three hundred. We should draw from what's there. Ritual, of course, is primarily how they educate the young. It's impossible to have a healthy educational system without ritual, and by ritual I don't mean forcing kids in to hear prayer from a pulpit. I mean imbibing the great myths and the great stories that are greater than industrial capitalism. In other words, our Creation stories.

These can be done through mythology and the very thing we've been talking about: praise. Maybe a more appropriate place than churches for teaching praise would actually be schools. Then maybe the elephants would be more relevant. As you point out, so much of what goes on in education is taking apart the elephant, getting the facts about the elephant. But seldom are we exposed to the awe of the elephant. That's what ritual can do. It seems to me that a condition sine qua non for education must be ritual, because we have to fill the hearts and minds with something very great. We must educate in awe again.

That's what myth is—it's a language for that which is too great to fit into mere factual compartments.

Then there's the problem of what kinds of ritual. Some children come from Christian backgrounds, some from Jewish, some have parents who are militant atheists, and so on. In Britain, schools are required by law to have an assembly period each day which is supposed to have a religious content—the opposite of the situation in the United States. But in most schools they don't know how to fill up

*the time. There is a void here, and also a great opportunity.
This opportunity is now greater than ever, because the financial cuts have meant that schools increasingly have to
call on parents and volunteers to help out. In our sons'
school in London, Jill now goes in regularly to lead the
assembly in chanting, and the children love it. This is one
very good way of cutting across religious divisions. But
when it comes to rituals, how can you do them without their
being utterly wishy-washy and devoid of content?*

That's where the creation story is bigger than any denomination. No one denomination owns this new one. That's
what binds us together. Atheists are just as subject to the
story of what happened and the fireball and the galaxies and
the sun as anyone else. It's a different way of teaching science. It would take mystical scientists like yourself to just
tell the creation story and then artists to help us retell it
through drama and participation—masks, costumes, multimedia, dances. That's what I mean by ritual.

I can see why we don't want to get into the area of particular, denominational rituals. We don't need to. The ritual
that the people have to hear on the planet today is about the
sacredness of the planet, how it got here, how we can preserve it and how we need to and how it's blessing us.
There's no limit of rituals: the story of the trees, the story of
the animals, the story of humans, the story of food, the story
of Gaia.

*But the problem is that neo-Darwinians have taken over
the creation story of life. If you told this story in a way that
involved any higher purpose, consciousness, mind, soul,
spirit—which is the only way a ritual would make sense—
they'd object straight away on the grounds that it's all a
matter of blind chance. They would accuse you of putting*

across anthropocentric and mystical thinking and cor-
rupting the young. How would you deal with that?

Physicists aren't so bad. The Big Bang is so remote that it
doesn't engage these emotions as much. Biology engages
them a great deal more because it's much closer to us.

But using the mentality upon which schools are built,
don't the biologists have to stay in the biology lab? If this
assembly is called religion, the biologists just stay away. Every religion has a creation story.

Yes, but I think most children wouldn't be very interested
in the creation of distant galaxies; after all, they've never
seen one. No part of their education involves looking at the
stars.

They must see at least one, that's the whole point! Bring
in a telescope, show them Saturn.

That's part of the reform of science teaching. In England,
I've been working with a group of science teachers trying to
think of better ways of teaching science. One of the things
that became obvious is that there has to be more direct
experience. At present they learn about the stars and planets
from books, not by looking at the night sky. By the time it's
dark all the teachers and the children have gone home.
There would have to be a night class, led by someone who
knows the stars and the planets well.

Begin with the experience. Like in Ernesto Cardinale's
analysis in his poem on the Universe. He says: "We can
argue about the reason of the Universe [and I would add
that we can argue about the origin of the Universe] and we
can argue about the meaning of the Universe—but we can-

not argue about the beauty of the Universe." That's what we can agree on. You begin with beauty, which is to say with awe and praise. You take the kids out to experience it.

When we were shown, many of us for the first time in our lives, through telescopes, Saturn vibrating and alive, everyone was talking about it. What a thrill! But to get this when you're very young—that's how science should begin. Then keep it alive through ritual. With the new technology we could have videos of Saturn.

Another example are the pictures of the organs of our bodies—to create rituals with these. Everyone has a body, but we're taught to be embarrassed by it. We can honor the heart and the liver and the lungs and so forth through ritual. Some people would dress up like a lung, we'd have scientists telling us how the lung operates, and then we'd have poets with poems to the lung and dances to the lung. This could go on easily, we've got lots of organs and this could go on all year long.

And it could be done in stages. In the first year or two you could start with the bladder and the rectum.

OK. That's their interest. Then around sixth grade you might get into the sexual organs. Our sexual education is so totally secularized, it's about pumps and pistons and do's and don'ts and stopping infections. But you put sexual education in a context of wonder and mysticism and point out that in fact every world religion honors sexuality in that context, then it's healthy. Your experience in India I'm sure confirmed that. We have the Song of Songs in the Bible and we pay no attention to it.

That's a good point, to go along at certain stages of children's development as to which organs are the most fascinating to them. And instead of pretending they aren't there, which is our usual parental response, to really go for them. Put them in the context of cosmology, which is spirituality. And wonder. We'd have much healthier people as a result, and our therapy costs would go way down at midlife. Maybe even other things like divorce and relationship issues would disappear, if people knew themselves better.

It would involve bringing in dimensions of experience that are currently ignored. Our older son goes to the local state primary school, and this last year the whole school was doing projects on water. They learned about the water cycle, dams, bridges, ships. They did pictures, made model waterwheels, and it was all quite good. But there was absolutely no spiritual or mythic element.

However, in the basement of the school, which is a large Edwardian building, there's a spring that is the source of the Fleet River. Most kids had never seen it, most didn't even know it was there. We'd seen it because we were interested in the sources of the underground rivers of London. It's a mysterious place. You go down some steps and then, behind a door that's normally locked, you look down into a large vaulted chamber with three or four feet of clear water at the bottom.

Jill and I asked the headmistress whether she'd shown the children the spring. She hadn't. They'd been taught all about water and the sources of water and they had a spring right under the building, but no one had thought of telling the children about it. Jill then had the idea of combining showing them the spring with chanting and a ceremony. The headmistress loved the idea. So at the end of the term we had a ceremony where the children chanted songs to the water spirits, and the well was lit with candles. We handed everyone a penny before the kids went down to visit it, including the teachers. They filed down by candlelight, after

*chanting in the assembly room, and each child had a turn to
stand in front of the well, and they each threw in a penny
and made a wish. This was a wonderful ceremony, very
moving for everyone who took part.*

And very memorable.

*Oddly enough, even the science teacher, who I thought
might be a problem, turned out to have a great respect for
this spring and was rather protective of it. But the spring
had never before been publicly acknowledged. This simple
ceremony brought in another dimension, but only after a
whole year of dealing with water with no mythic or spiritual
dimension at all.*

*Small things like this are possible, but they need to be
integrated more with the teaching, and it's difficult to see
how they can be brought into science curriculums as currently taught.*

That's where it falls into the purview of religion. You
have this assembly or empty space because religion appar-
ently has nothing to say to people today and knows it. At
least, it stays out of the way. But there's a vacuum to move
into. It took the instinct of Jill to do that.

I remember a few years ago I was doing a workshop at
Evanston, Illinois, at Garret Theological Seminary, a Meth-
odist seminary. I had the participants doing circle dances
and chanting. A woman came up to me afterward and said,
"I like chanting so much more than singing because when I
was a little girl they told me I couldn't sing. So when I go to
church I don't feel at home at all because everybody's singing and I just peep. But here in chanting I can really let it
out."

Everybody can chant. If you've still got breath in you,

you can chant. Chanting is so much more nonelitist than singing, and it takes you away from the eye work. With singing you have to have the notes. Chanting is very different. It's something we carry within us. The idea of getting back again to chanting is primal to any ritual experience. We need to get back to these simple forms of expression and incorporate them. I heard that the children used to clap when Jill came into the room because she represented chanting and they were ready for it.

We're not talking about forcing something on people. We're talking about appealing to human nature. There's a delight in ceremony, ritual, drama. All kids, especially teenagers, love to be hams, they love to act. They don't get a chance, because they're watching television instead of acting. So the whole idea is for the children in these schools to create their own programs, their own news. Generating creativity, that's what's missing in these schools. The perfect place for it in the system you're describing is religion.

Healthy religion did things like that once. You can look at Chartres cathedral and see the stained-glass windows, which were the color television of the day, and realize something interesting was going on. There's a precedent for this; there's a history.

In New York City a year or two ago, all art classes were canceled in the high school system because of financial difficulty. The result was this: the art community of the city said, "This is intolerable." So they organized and started raising money outside the school system, which gave them total independence. They now offer their services to these schools. But they do art very differently now because they're not being controlled by the schools. And frankly, I think if they were instructed just a little bit they could be moving the schools with ritual and with art as meditation.

The collapse of what's going on around us, including education, is a potential breakthrough. I think we should pay much more attention to this. It's a real opportunity to start infiltrating with a broader definition of what learning is, which includes body and the whole mind.

Part of a traditional education is apprenticeship, which in India is raised to the highest form in the ashram system, where the student goes to live with the teacher, the guru. In families within the occupational castes, like potters and weavers, the children are automatically living with people who engage in an occupation: they see their parents doing their work, and from an early age the children help. Doing useful work is part of childhood. Our children are economically redundant until they're grown up. But the idea that children do useful work is part of most traditional cultures.

In the work is the learning.

Apprenticeship is the form in which this traditional system was adapted in the West, and it still exists in the learning of trades like plumbing and carpentry. The best way to learn to be a plumber is to work with one.

This principle could well be extended to cover subjects normally taught in an academic way in classrooms. For example, the best way to learn about plants is not through schoolbooks, but through gardening. You really learn about plants if you grow them outdoors. There are plenty of accomplished gardeners around, including retired people who've had years of experience and now have time to spare. Not all children are interested in gardening, but some are, and for those who are, setting up a system whereby people who are good at gardening could have children work with them and maybe have their own patch of garden would be a tremendous education in botany.

In fact, even conventional science teaching involves a great deal of practical work in a laboratory; that's one of its strengths, because learning science does involve a kind of apprenticeship. The trouble is that a lot of what you do is not making or growing things, but taking them apart. If you do botany at school or college, you don't usually start by growing plants, you start by dissecting them.

Anything that's practical, involving learning by doing, you can learn by apprenticeship. There's a tremendous overemphasis in our educational system on reading and writing and merely mental skills.

That's the key: hands-on. What we create ourselves, what we make ourselves we remember. For example, my brother who's a very creative teacher, years ago had fifth and sixth graders, and for physiology he had them make skeletons. He got a call from a parent who said, "I'm a professor of medicine and I just quizzed my daughter. She knows more about the human body than my medical students do after their first year of medicine. What are you doing in class?" "Well," my brother said, "we made skeletons."

To make a skeleton you've got to know something. For his class on history, he turned the classroom into Chartres cathedral and he had a team making flying buttresses and a team doing stained glass and a team doing pillars and so forth. He did this experiment without telling the kids: he brought in a stack of books on flying buttresses and didn't say anything—just left them on the table. The team doing flying buttresses had to learn about them and they saw the books, took them, devoured them. Two weeks later my brother quizzed them without telling them he was going to do it. He said that he was convinced these kids knew more about flying buttresses than any engineer in the country. His whole thesis was that a child's mind will absorb anything if it's a delightful project. If you have the responsibility to make something out of your knowledge, then it's going to happen.

Part of apprenticeship is the mentor relationship, but part of it is hands-on experience. We have to ask, How can more of our knowledge be translated into hands-on things? That history class where the kids made Chartres cathedral—there's a method here, what I call art as meditation. It's all hands-on. Massage, for example, teaches you the body and if it's done in a context of spirituality, it also teaches you the sacred. All art is hands-on. That is why art as meditation must be the core of any curriculum.

When you think about it, isn't it possible that all knowledge in some way is art and this implies hands-on? It all relates to making or creating in some fashion, and we don't honor that enough.

CONCLUSION

In the Preface we spoke of recovering "a sense of the sacred" and of how this can address the despair and disempowerment that so many people feel in our time. Do these discussions on Nature and creation spirituality, grace and praise, soul, prayer, darkness, morphic resonance and ritual, and revitalizing education in fact assist us in bringing back a sense of the sacred? Did they do so for us?

Yes. One way in which they did so was by showing how the insights that science has given us have been kept separate from the realms of intuition and imagination. Much of what science has revealed about Nature has been left in a sterile state, insulated from the world of the spirit. With the riches of Nature that science has opened up come new opportunities for thanksgiving, for praise, and for wonder at the creativity underlying all things. In this fuller context the

truth of the mind and the truth of the heart can come together. Effective praise depends on paying attention, and
science makes available to us both a new image of the creative, evolving cosmos and an amazing wealth of detail
through which our praise can be informed. Moreover, a
shared cosmology provides the basis for a shared morality.
Shared wonder and awe at creation provides a substratum
for genuine community. The souls of the young expand
when they learn that their lives are not trivial but are part of
an ancient story.

For these reasons, these dialogues have made us more
hopeful. A new vision of our past as well as our future lifts
us up and fills our imaginations with new possibilities. Hope
is built on what is possible, despair on what is impossible.

We are aware that these dialogues are only a beginning in
the new convergence of science and spirituality. We trust
that they will help to stimulate others and to awaken hope in
them as they have in us.

BIBLIOGRAPHY

Berry, T., and B. Swimme. *The Universe Story*. San Francisco: Harper, 1994.

Cardenal, E. *Cosmic Canticle*. Willimantic, CT: Curbstone Press, 1993.

Fox, M. *On Becoming a Musical Mystical Bear*. New York: Paulist Press, 1972.

Fox, M. *Breakthrough: Meister Eckhart's Creation Spirituality in New Translation*. New York: Doubleday Image Books, 1980.

Fox, M. *Original Blessing: A Primer in Creation Spirituality*. Santa Fe: Bear and Co., 1983.

Fox, M. *The Coming of the Cosmic Christ*. San Francisco: Harper and Row, 1988.

Fox, M. *Sheer Joy: Conversations with Thomas Aquinas on Creation Spirituality*. San Francisco: Harper, 1992.

Fox, M. *Confessions: The Making of a Post-Denominational Priest*. San Francisco: Harper, 1996.

Hildegard of Bingen. *Illuminations of Hildegard of Bingen*. Santa Fe: Bear and Co., 1985.

Kuhn, T. S. *The Structure of Scientific Revolutions* (2nd ed.). Chicago: University of Chicago Press, 1970.

Mowinckel, S., in: C. Westermann's *Blessing in the Bible and the Life of the Church*. Philadelphia: Fortress Press, 1978.

Piaget, J. *The Child's Conception of the World.* London: Granada, 1973.

Rifkin, J. *Beyond Beef: The Rise and Fall of the Cattle Culture.* New York: Dutton, 1992.

Rumi. *A Garden Beyond Paradise: The Mystical Poetry of Rumi.* Translated by J. Star and S. Shiva. New York: Bantam Books, 1992.

Sheldrake, R. *A New Science of Life: The Hypothesis of Formative Causation.* Los Angeles: Tarcher, 1982.

Sheldrake, R. *The Presence of the Past: Morphic Resonance and the Habits of Nature.* New York: Times Books, 1988.

Sheldrake, R. *The Rebirth of Nature: The Greening of Science and God.* New York: Bantam Books, 1990.

Sheldrake, R. *Seven Experiments that Could Change the World: A Do-It-Yourself Guide to Revolutionary Science.* New York: Riverhead Books, 1995.

Spretnak, C. *States of Grace: The Recovery of Meaning in the Postmodern Age.* San Francisco: Harper, 1991.

Teresa of Ávila. *The Interior Castle.* New York: Paulist Press, 1979.

Whyte, L. L. *The Unconscious Before Freud.* London: Friedmann, 1979.